Air Pollution and the Social Sciences

edited by
Paul B. Downing

The Praeger Special Studies program—utilizing the most modern and efficient book production techniques and a selective worldwide distribution network—makes available to the academic, government, and business communities significant, timely research in U.S. and international economic, social, and political development.

Air Pollution and the Social Sciences
Formulating and Implementing Control Programs

PRAEGER SPECIAL STUDIES IN U.S. ECONOMIC AND SOCIAL DEVELOPMENT

Praeger Publishers   New York   Washington   London

PRAEGER PUBLISHERS
111 Fourth Avenue, New York, N.Y. 10003, U.S.A.
5, Cromwell Place, London S.W.7, England

Published in the United States of America in 1971
by Praeger Publishers, Inc.

All rights reserved

© 1971 by Praeger Publishers, Inc.

Library of Congress Catalog Card Number: 75-153390

Printed in the United States of America

This book is dedicated to the men and women of the various air pollution control agencies in the hope that the information provided here and the additional research it might stimulate will help them do their jobs more effectively.

PREFACE

The contents of this book constitutes a review of the literature in the social sciences dealing with air quality and related environmental quality issues. The period covered extends to mid-1970. The disciplines covered include sociology, psychology, political science, law, and economics. A combined bibliography is provided to facilitate further research. It is the hope of the authors that this work will stimulate others in their research.

Each of the five main chapters is an attempt to use the available literature to determine what each discipline can contribute to solving the air pollution problem. In this effort, gaps in the literature are noted, and research is recommended to fill these gaps.

As the reader will note, the chapters are not merely a review of who said what. Rather, they include careful and thorough synthesis of the issues involved and important new insights from each author.

The authors' general perspective is that the air pollution problem is of major concern and that progress toward a solution is an immediate goal. Furthermore, the authors take the position that the problem will not be solved without a substantial contribution from the social sciences. In fact, some feel that the solving of the technical problems of air pollution control will prove to be relatively insignificant in moving toward a solution to the problem. It is the social, political, legal, and economic problems that are currently preventing a solution.

This volume was originally written during the spring and summer of 1970 for the University of California's Project Clean Air and revised during the winter of 1970-71. Project Clean Air is the University's applied research program in air pollution control. This study is published with the permission of the Board of Regents of the University of California. We wish to express our thanks for this permission and for the University's financial support for this effort.

Our format consists of an introductory chapter which attempts to define the central issues of interest to the social sciences, five chapters by authors in five different social science disciplines, and a concluding chapter which attempts to use the information presented in the first six chapters and the technical and health information presented in the seven taskforce assessments of Project Clean Air to produce a model for policy analysis.

The editor wishes to thank each author for his contribution, as well as Lytton Stoddard for his assistance in several aspects of the preparation of this book.

A NOTE ON LITERATURE CITATIONS

There is a large volume of literature cited in the seven chapters of this book. In order to conserve on space, the literature citations refer to the extensive bibliography compiled by Michael L. Fox which is found at the end of this book, and a modified "date-finder" system is employed. Each citation in the text consists of the author's name, date of publication, and page number, if necessary. Thus, the citation (Gold, 1970: 25) refers to a work by David Gold, published in 1970, and, specifically, to page 25. The full citation to this work is given in the sociology section of the Bibliography, under Gold. The key to what section of the Bibliography to consult is the subject of the chapter in which the citation is given.

CONTENTS

	Page
PREFACE	vii
A NOTE ON LITERATURE CITATIONS	ix
LIST OF ABBREVIATIONS	xv

Chapter

1 AN INTRODUCTION TO THE PROBLEM OF
  AIR QUALITY
  Paul B. Downing   3

    Definition of the Air Pollution Problem   3
    Why the Problem Remains   4
       Lack of Effective Polluter Action   4
       Lack of Effective Public Action   5
       Lack of Effective Government Control   6
       Growing Public Interest   7
    The Nature of the Solution   8
       The Level of Control   8
       Technology and Control Costs   8
       Reallocation of Resources   9
       Distribution of Financial Burden   13
    Conclusion   14

2 AIR POLLUTION AS A PROBLEM FOR
  SOCIOLOGICAL RESEARCH
  Harvey Molotch and Ross Charles Follett   15

    Introduction   15
    Issues and Analysis   15
       The Heritage of Human Ecology   15
       The Theory of Cultural Lag   17
       The Questioning of Accumulation   19
       What People Want   20
       Clean Air as a Social Movement   24
       Clean Air and Community Organization   26
       Air Pollution and the Structure of Power   28
       Air Pollution and the Mobilization of Bias   30

Chapter		Page
	The Administration of Solutions	32
	Tasks Recommended for Air Pollution Research	34
3	**PSYCHOLOGICAL AND BEHAVIORAL EFFECTS OF AIR POLLUTION** Robert W. Reynolds	37
	Introduction	37
	Issues and Analysis	37
	Studies in the Soviet Union	38
	Studies in the United States	40
	General Discussion	42
	Tasks Recommended for Air Pollution Research	44
4	**POLITICAL SCIENCE AND AIR POLLUTION: A REVIEW AND ASSESSMENT OF THE LITERATURE** Ronald O. Loveridge	45
	Introduction	45
	Issues and Analysis	46
	Political Science and Air Pollution Research	46
	Politics of Air Pollution	49
	Research and Air Pollution	65
	The Public	65
	Interest Groups	74
	Policy Legitimation	74
	Policy Implementation	75
	Research Prospects and Projects	78
	Notes	80
5	**AIR POLLUTION AND LEGAL INSTITUTIONS: AN OVERVIEW** James E. Krier	87
	Introduction	87
	Legal Perspectives	90
	The Nature of the Problem	90
	The Problem and the Political Process	95
	Air Pollution and the Courts	99
	Objectives of Environmental Litigation	99
	Shortcomings and Palliatives	106
	Pollution-Control Legislation and Administration	111
	Regulation	112
	Subsidies	116
	Pricing	118

Chapter	Page
Conclusion	121
Research Recommendations	122
Notes	123

6 THE ECONOMICS OF AIR POLLUTION: A
  LITERATURE ASSESSMENT
  Robert J. Anderson Jr. and Thomas D. Crocker — 133

Economic Theory and Air Pollution	133
Indivisibility and the Market	134
The Central Role of Extramarket Institutions	136
The Potential Role of Economics in Air Pollution Control	138
A Reader's Guide and a Caveat	139
Theoretical and Empirical Studies of the Emitter	140
Engineering Studies	140
Survey Studies	141
Engineering-Survey Studies	142
Some Neglected Aspects of Emitter Behavior	142
Theoretical and Empirical Studies of the Receptor	143
Political Studies	144
Interview Studies	145
Response-Surface Studies	145
Market-Behavior Studies	148
Aggregated Systems	149
Control Instruments	151
Importance of Property Rights	151
Customary Discussions	152
Assessing Results of Instrument Implementation	154
A Prelude to Future Research	157
Notes	160

7 CONCLUSION: A POLICY TRADE-OFF ANALYSIS
  MODEL FOR AN AIRSHED
  Paul B. Downing — 167

Introduction	167
Basic Issues and Suggested Decisionmaking Methodology	168
The Level of Air Quality	169
How to Obtain Desired Air Quality	171
Interdependency of Decisions	171
Least Cost, Air-Quality Sets	171
Method of Solution	174
Related Secondary Effects	175

Chapter	Page
Benefit Sets	176
Administrative and Institutional Structure of Control Efforts	178
Interrelationships Among Components of the Model	183
The Desired Level of Quantification of the Model	184

AN AIR POLLUTION BIBLIOGRAPHY FOR THE SOCIAL SCIENCES
Michael L. Fox — 189

Economics	189
Law	209
Planning	229
Political Science	235
Psychology	244
Sociology	253
General	263

ABOUT THE EDITOR AND THE CONTRIBUTORS — 271

LIST OF ABBREVIATIONS

CAN	Clean Air Now
CO	Carbon monoxide
CRF	Continuous reinforcement
DRL	Differential Reinforcement of Long Latency Responses
EPA	Environmental Protection Agency
GASP	Group Against Smog Pollution
HC	Hydrocarbons
HEW	U.S. Department of Health, Education, and Welfare
ICP	Informational, contractual, and policing (costs)
MVPCB	Motor Vehicle Pollution Control Board
NAPCA	National Air Pollution Control Association
NO_x	Nitrogen oxide
PAN	Peroxyacyl nitrate
PB_zN	Peroxybenzoyl nitrate
ppm	Parts per million parts (of effluent release)
SOS	Stamp Out Smog
SO_2	Sulfurdioxide
SST	Supersonic Transport

Air Pollution and the Social Sciences

CHAPTER

1

**AN INTRODUCTION
TO THE PROBLEM
OF AIR
QUALITY**

Paul B. Downing

The intent of this chapter is to briefly introduce the reader to some of the more important issues discussed in the main body of this work. The framework provided here will help relate the points made by each writer to the central issues of air-quality and its control. We found such a framework helpful in our efforts to understand the full range of implications of each chapter as well as agreements and differences among the writers. The reader should be warned that this author is not an economist, and the discussion presented here is likely to reflect our biases. However, we have tried to fairly represent each of the disciplines concerned with the problem.

In this chapter, we first attempt to define the air pollution problem. Next, we suggest some of the reasons why the problem remains. We then turn to a brief discussion of the nature of the solution, or at least where social scientists might look for solutions.

DEFINITION OF THE AIR POLLUTION PROBLEM

Society would prefer a clean-air environment to a polluted one. However, the attainment of clean air requires an expenditure of

---

Portions of this chapter are based on material originally published as "Solving the Air Pollution Problem: A Social Scientist's Perspective," Natural Resources Journal, 1972. Permission to use this material is gratefully acknowledged.

resources, and society has other uses for these resources. In the United States, the resource-allocation process is left to the market mechanism except when society is not satisfied with the result. In such cases, society, through government, interferes with the market.

What, then, is the air pollution problem? Briefly stated, society appears to be demanding a better air quality than it is currently getting in the market. That is, society is willing to allocate more resources to the production of clean air than the market is now allocating to that goal. Having said this, let us retract somewhat from this stand. Air quality is not a problem of universal concern throughout the United States. People in many parts of the country are not now suffering from sufficiently poor quality air to justify additional expenditures on control. In other areas, the air may be of relatively poor quality, but the people of the area may prefer to continue breathing that level of air quality rather than taking resources away from other activities which they regard as more important. In other words, the choice of the quality of air is a local choice which depends on the local meteorological conditions, mix of pollutants released, technology and cost of controls, and the local preferences for clean air relative to other individual and social goals. Throughout the remainder of this chapter, and, for the most part, in the other chapters, it is assumed that society desires a reallocation of resources toward the production of better air quality.

WHY THE PROBLEM REMAINS

A combination of factors contribute to the lack of a solution to the problem. Many of these factors are technological in nature. Concern here will be exclusively with the institutional factors which cause an allocation of resources away from air pollution control.

Lack of Effective Polluter Action

A rational firm tries to produce its goods at the lowest possible cost. One way to cut costs is to release unwanted byproducts into the environment. In this way, the firm saves itself the cost of collecting and removing these byproducts from the place of production. The release of these wastes into the atmosphere causes others to suffer losses (poor health, loss of view due to haze, and so forth), for which they receive no compensation from the firm. There is little incentive in the private market for a firm to reduce the losses it imposes on others. Under our legal and economic structure, society often does not provide the incentive to reduce the losses that a firm imposes on others. The net result is a situation where the firm pays less than

THE PROBLEM OF AIR QUALITY

the true cost of its production (where the true cost of production includes the firm's out-of-pocket costs and the losses it imposes on others). The lower costs are reflected in the lower price charged for the firm's products. This causes the firm to produce and sell more than would be the case if it paid the full costs. Thus, more pollution than is desirable is produced and released.

It is in the interest of each firm to resist any government control actions which would increase its out-of-pocket costs of production. Furthermore, some industries are more adversely affected by control efforts than others. These industries also tend to have few firms and large sales volumes. For example, they include automobile manufacturers, oil companies, and electric power utilities. This combination of general resistance to control efforts by all firms and strong resistance by the relatively few but politically and economically powerful firms most adversely affected results in strong anticontrol lobbies at all levels of government.

The individual suffers from pollution. However, he is also a polluter. He uses a car that emits pollutants. He burns trash in his backyard or does so indirectly through local government as his agent. The reasons he does not take action to reduce his pollution of the atmosphere are in part the same as those cited for the firm. Pollution control costs money or loss in time available for other pursuits, and there is little incentive to expend the effort. In addition, the individual perceives the problem as very large and his contribution to the problem as too small to have any effect on the quality of the air he breathes. This combination of cost without significant, perceivable, personal benefits results in a passive, noncooperative attitude toward control efforts. Typical comments we hear when we ask friends about air pollution control are: "I'm not going to do anything unless everybody else does too," and, "Somebody (the government) ought to do something about it."

Lack of Effective Public Action

The damages caused by air pollution accrue to all individuals living in the area of the release. Similarly, the benefits of improved air quality accrue to all individuals in the controlled area. Thus, successful efforts to control pollution by some individuals would cause others who did not join the effort to reap benefits as well. This provides a strong incentive for many to refuse to help in attaining control of air pollution because each individual knows he will benefit from a successful effort without any expenditure of his own resources. This is a classic example of the "free-rider" principle.

At the same time, each individual compares his power to affect governmental control decisions with that of the anticontrol faction. In this comparison, he feels impotent. The result of this combination of free-rider problems and a perceived lack of political power has been a relatively weak (or nonexistent) procontrol lobby.

Lack of Effective Government Control

While it is true that some government agencies now in operation have slowed the increase in air pollution levels, and, in some cases, may even have reduced pollution levels, people still complain of dirty air. The combination of a strong anticontrol lobby and a weak procontrol lobby have understandably resulted in a relatively weak control effort. This weak control effort may be due in part to poor management by control-agency officials, but a greater share of the blame can be placed on the organizational structure of the control effort. Public opinion has forced the passage of bills to control pollution. However, there is more than one way to subvert control efforts. Thus, we find that agencies already in existence lack adequate control authority because of diversified control responsibilities. For example, in California, the Air Resources Board has control responsibility over automobile emissions and county Air Pollution Control Districts have control responsibility over stationary sources in their respective counties. The Los Angeles Air Basin includes parts of six counties, each with different jurisdictional organizations and regulations for air pollution control.

A second factor that leads to lack of effective control is the limited amount of resources control agencies are provided with to do their job. A typical political maneuver is to approve a program but to provide the program with little or no funding so that it cannot really do the job it is assigned to do. With this tactic, the politician can point to his stand against air pollution when questioned by the electorate and yet not alienate the anticontrol lobby which may support his re-election.

A third factor that leads to lack of effective control is the difficulty in enforcing controls under current forms of regulation. Standards are usually enforced by a permit or certificate issued to the firm. The cost of administering such a system is very high, and, in light of the limited resources available to control agencies, is critical. Furthermore, the inspection process often interferes with the normal operation of the plant. As stated in Chapter 5:

[Noncompliance is] enforced by criminal process, probably the most cumbersome coersive tool we have. The violator is protected by all the constitutional protections which apply to any criminal trial. He can demand a trial by jury and unanimous verdict (and this against the heavy burden of proof faced by the prosecution).

Even if the firm is convicted of a violation, the penalty imposed is usually token.

Finally, control-agency officials are sensitive to political pressures, either explicit or implied, from the anticontrol lobby, and conceivably from the procontrol lobby. The agency is dependent on political entities such as state legislatures for annual appropriations. Its officials are aware of any action that might offend the political powers which control the resources needed for their control program. This can result in lax administration of control regulations or appreciable variances for large and politically powerful firms. The firms themselves use this fact in their relationship with the control agency. The typical threat is that controls imposed by the agency would cost too much, and the firm would have to close its plant and move. If this were true, the plant would be marginal, and it is likely that the firm would either rebuild or move shortly even without additional air pollution controls being placed on it. In fact, as one study points out:

[In] Los Angeles, not a single industrial establishment, large or small, old or new, has been forced to move away or go out of business because of the cost of complying with air pollution regulations. [Faltermayer, 1968: 73]

Thus, we find strong pressures placed on government agencies to go slow on control efforts and weak pressures placed on them for stringent controls. The result is obvious. One of the jobs of the social sciences is to find ways of neutralizing this anticontrol bias in our legal, governmental, and socioeconomic structure.

Growing Public Interest

In recent years, public interest in improving environmental quality has grown rapidly. It is now a leading political issue. This increasing interest stems from a combination of economic and sociological factors. For many families in the middle, upper-middle, and upper-income classes in the United States, income has increased to the point where the acquisition of additional material goods no longer serves as the primary personal goal. The following is noted in Chapter 2:

Among those Americans who have lived with their basic

"needs" of food, clothing, and shelter satisfied, there is
emerging a new consciousness of alternative goals. . . .
among them would be a physical environment which is
healthful and aesthetically pleasing.

This, coupled with the growing availability of leisure time, adds power to the observation.

THE NATURE OF THE SOLUTION

There are four aspects of the solution which are discussed below: The desired level of control, the technology and cost of control, the attainment of resource reallocation, and the distribution of the financial burden of control.

The Level of Control

The level of control of air pollution that is socially desirable is a function of the cost of control, the technological effectiveness of control measures, and the damages averted by lowering pollution levels. These are the elements of the traditional benefit-cost analysis. The decision rule is to expand air pollution control to the point where the additional cost of one more unit of control (marginal cost) equals the value of the additional damages averted (marginal benefit). This equality defines the point where total net benefits (total benefits less total costs) are a maximum and hence the optimal allocation of resources to air pollution control. If we accept this criterion, the task remains to quantify these costs and benefits. While the quantification of benefits is conceptually easy, in practice, current methodological tools can quantify only some of these benefits. Many others remain unquantified or, in some cases, quantified in terms that are not commensurate with the dollar measure used for costs. This leads to difficulties in the application of the technique. An important point which must be stressed is that this methodology leads one to conclude, due to the cost of control and the low levels of damages at high control levels, that there will be some nonzero level of pollution desired by society.

Technology and Control Costs

Some people claim that most of the technology necessary to "solve" the air pollution problem is already available (Faltermayer, 1968: 85). While it is true that substantial improvements in air quality could be made if existing control devices were universally installed, many technical problems would remain. In the next few years, much work is needed in the development of devices which do an adequate pollutant-removal job inexpensively and reliably. In the long run, growing population will dictate higher levels of control. The attainment of current and future air-quality goals is not going to be technologically easy. For example, the California Air Resources Board calculated:

> Current maximum hourly oxidant averages in the San Francisco region are about 0.35 ppm, and in the Los Angeles region about 0.75 ppm. Based on current auto exhaust control programs, including the standard adopted by the Air Resources Board for 1975, the maximum hourly average oxidant values in Los Angeles will probably not be below 0.3 ppm by 1980, will not be below 0.2 ppm by 1985, and cannot reach 0.1 in 1985 (the state standard). For the San Francisco region similar estimates indicate that the oxidant values will be about 0.15 in 1980, and will reach 0.1 in 1985. [Technical Advisory Committee, 1970: 12

With all the research that has been completed to date, it seems unlikely that a technological panacea will be discovered. Society cannot afford to wait and hope that such a panacea will appear. We should more adequately tackle the job of adapting and implementing the technology available now because damages from air pollution are occurring now.

Reallocation of Resources

At the present time, the U.S. economy is using its resources to produce goods, and the secondary effects of this production on the environment are substantially ignored. As was pointed out above, there is good reason for this attitude among individuals and firms. The legal and social structure is such that some form of government action is necessary to force individuals and firms to consider in their resource-use decisions the affect of their actions on the environment and hence overcome the anticontrol bias. The goal is to reallocate the use of the nation's resources away from pollution and toward control. This goal assumes that we now have too much pollution, an assumption which seems reasonable in light of the growing public concern with the quality of the environment.

Resource allocation for air pollution control is a function of the desires of society, the method by which these desires are translated into group (political) action, and the effect of political action on the structure of air pollution control institutions and on the resources available for control efforts.

The first issue, then, is who is in charge of air pollution control and is that person or institution responsive to all sides of the issue? As was suggested above, those presently in charge (politicians as well as state and local air pollution control officials), are often not responsive to all sides of the issue. Rather, they are biased by the system as it now exists to be less effective than society might like. We must find ways to design institutions that fairly adjudicate among all sides of the control controversy.

The second issue is the method by which we can bring about reallocation desired by the decisionmaking institution. There are five basic forms of control instruments that are either in use or that could be used for air pollution control. They include prescriptive regulations, effluent standards, air-quality standards, pricing and taxing systems, and property-rights restructuring. It is unlikely that the use of any one form of control to the exclusion of the others will succeed in attaining the quality of air the public desires. Each form has advantages and disadvantages.

Prescriptive Regulations

These regulations state that a specified action must or must not be taken. The requirement that all used automobiles sold in the state of California have an crankcase breather device installed and in good operating condition is an example of such a regulation. The major advantage of this form of regulation is its relative ease of enforcement. A service-station mechanic can inspect the automobile in a few minutes and issue a certificate of compliance. This involves very little time lost by the purchaser, and the Motor Vehicle Department only has to check to make sure the certificate is valid. The major disadvantage of prescriptive regulations is that they are inflexible. If a new device were developed which controlled hydrocarbon emissions more effectively for less money and with no increase in the emission of other pollutants, its substitution for the breather device in future installations would require legislative action. And such action is likely to be slow in coming because of the vested interests of those who benefit from the old regulations.

Effluent Standards

These regulations state that a particular source of pollution must control its emission of identified pollutants to a specified level. The specification can be in terms of parts per million parts of effluent release (ppm) or some measure of the total amount of pollutants released, such as the grams per mile standards for automobiles. The total release form (if properly measured) is preferable because the

THE PROBLEM OF AIR QUALITY

ppm standard stimulates dilution rather than control. Thus, the California effluent standards for 1970 automobiles stated that an automobile could not emit more than 2.2 grams per mile of hydrocarbons. The major advantages of the effluent standard are its relative ease of enforcement (at least in theory) and its freedom from specification as to how the standard is to be met. Enforcement is theoretically easy because all that is necessary is a periodic check to insure that the emitter is not violating the standard. However, there are two problems with enforcement. One is the difficulty of measuring the effluent. Access to the effluent must generally be obtained, and this can only be done with the emitter's knowledge. Even when access is obtained, there is still the problem of measuring the pollutants accurately and inexpensively, a technological problem that in many cases has not been solved. The second difficulty is that there are large numbers of emitters which should be monitored. With present technology, this becomes a time consuming, expensive, and difficult task.

The effluent standard is stated in a way that allows each polluter to determine the best method for him to meet the standard. For one firm, this may be through installation of a control device such as a stack-gas scrubber; for another, it may be through use of a different fuel. A third firm might find a change in its production process to be the least expensive method of meeting the same standard. Some find the standard difficult and expensive to meet. They often apply for and are granted a variance on the grounds of hardship. Other emitters find the standard easy to meet at little or no expense. Yet, they are not asked to meet stricter standards. Stated another way, some firms find pollution control expensive while others find it relatively cheap. The same level of pollution control could be obtained for a smaller cost to society if the latter were to control emissions to a greater degree than the former, rather than having both control to the same degree.

Air-Quality Standards

These regulations state a specific level for pollutants in the air that cannot be exceeded. In California, for example, the air cannot exceed 0.10 ppm oxidant for more than one hour. The major advantage of this form of regulation is that it explicitly recognizes the goal of air pollution control, improvement in the quality of the ambient air. The standards are typically based on consideration of technical feasibility and health effects. The major disadvantage of air-quality standards is that they do not solve the problem as to which polluters are to control how much of their emissions. Such standards put a large burden on control agencies who must determine the best way to meet the standards. Air-quality standards do not stand by themselves. They must be supplemented by other forms of regulation. Thus, air-quality standards are often treated as goals to be strived for, rather than as legal constraints that must be met.

Pricing or Taxing Regulations

These regulations impose a charge or tax for the release of pollutants into the atmosphere. The charge would vary with the type of pollutant released, the amount released, and the time of release. The basic principle is that if a polluter releases pollutants which cause damage, he must pay society to compensate for these damages. If he releases less pollutants and thus causes less damages, he will be required to pay less. Thus, an incentive is provided for private action to reduce pollution in order to avoid or diminish the charge. A subsidy may be employed in the same way. In this case, the higher a firm's control efforts, the higher the subsidy it receives. However, there are two major problems with subsidies that do not exist with charges. First, it is difficult to determine if a firm might install the control device even without the subsidy. Second, a subsidy may offend the public's sense of equity in that people are paid to do what their social conscience should require them to do.

The charge system's major advantage is that it provides for maximum flexibility in individual emission-control efforts. Each emitter balances savings in pollution charges against his cost of control. He minimizes the sum of his control costs and pollution charges by equating the marginal savings of pollution charges with the marginal cost of additional control necessary to bring about the reduction in charges. This level of control is likely to be different for each polluter even if the emitters are engaged in the same activity (e.g., power production). The over-all level of control will be determined by the level of the charges. The major disadvantages of this system are its administrative problems and the related measurement problems of applying the charges. In order to equitably apply a charge system that has some sophistication in terms of variations with quantities of pollutants released and their time of release, some fairly accurate, periodic, and/or continuous measurement of emissions must be made by the regulating authority. This must be done for each emitter. As in the case of the effluent standard, the measurement problem may, in some cases, be impossible with the instrumentation now available. In addition, the computing of charges and their collection can add substantially to administrative costs.

Restructuring Property Rights

This action would vest the right to use the air in either the recipient of the pollution or some central authority, rather than the polluter, as is now the case. The major advantage of this action is

that it would provide a bias toward control in the legal system that
could be used to offset the bias toward pollution which now exists.
Currently, a party who has been damaged by air pollution must prove
in court that emitter A damaged him. He must establish that he was
damaged and that emitter A did it and not emitter B. This is almost
always an impossible task. Under restructuring, emitter A could not
pollute if he damaged anyone excessively, and the burden of proof
would be on him rather than on the pollution recipient. Although little
work has been done on the possibility and results of the restructuring
of property rights, one presumed disadvantage is that it still would
not provide a marketable property right in the same sense as land
ownership is a marketable property right. Instead, it substitutes untransferable rights with a bias against pollution for the current, untransferable rights with a bias toward pollution. To get around this
problem, various licenses or permit systems have been suggested.

The permit-system concept was suggested originally by Crocker
in Wolozin (1966) and further developed by Dales (1968). Permits to
pollute the air would be sold at auction by the air pollution control
agency. The total sum of pollution to be allowed in the airshed (as
determined by some government agency) would determine the total
number of permits to be sold. Both the public and the emitters (manufacturers, and so forth) could bid for permits. The public, as represented by conservation groups and perhaps governmental agencies,
would purchase permits in order to deny them to the emitters and
thereby reduce pollution, while the emitters would buy them in order
to avoid control costs (or perhaps in order to operate). This system
would provide each side with equal access to the air resource and, in
theory, could work well. The problem, however, is that the public
would again try to act as free riders. For example: "If a conservation
group were to buy up pollution rights, I could enjoy the cleaner air
without compensating the group." Much additional study is needed on
the subject of property-rights restructuring.

Combinations of Control Instruments

It should be obvious from the above that each control instrument
has its advantages and disadvantages, and no single instrument will
do the job perfectly. While it is probably also true that no combination
of control instruments will be perfect, deficiencies in one instrument
can be offset by strengths in another. Thus, a combination system is
likely to be the best control method.

Distribution of Financial Burden

One of the most important issues that must be faced in air pollution control efforts and the one around which most of the control controversies center is: Who pays for control? Surprisingly, economists have little they can say on this subject, even though it would seem to be at the heart of their discipline. Instead, they say that resource allocations will be different depending on who pays for control, but there is no way to judge, using economic criteria, which is the better allocation. Instead, the issue becomes one of equity (and, in our system, politics). The polluter may be forced to pay. This seems equitable to the general public: The polluter causes the problem, so he should be the one who pays for eliminating it. However, political constraints limit the use of this concept. The recipient of the pollution may pay for control because he benefits directly from cleaner air. Politicians may find this alternative attractive because each of the many recipients would be charged a relatively small amount per year rather than a few polluters who would each pay a large amount. A third alternative is to have all taxpayers finance control. In an effort to spread the burden over as many people as possible, it is likely that politicians will use all three sources of funds.

CONCLUSION

In this chapter, we have argued that the air pollution problem is one of insufficient allocation of society's resources toward air pollution control. We have also suggested why this has occurred. Finally, we have suggested several of the more important aspects as to where social scientists would look for solutions to this problem. The remainder of this book is devoted to the exploration of these issues and what social scientists can now say about them. All the answers are not known, but to the extent that they are able, the authors attempt to provide answers. Where they are uncertain of the answers, further research is suggested.

CHAPTER 2

AIR POLLUTION AS A PROBLEM FOR SOCIOLOGICAL RESEARCH

Harvey Molotch
Ross Charles Follett

INTRODUCTION

Given the fact that there is a near dearth of explicit sociological research on the subject of air pollution, this chapter represents an attempt to define those tasks which could serve as a prelude to the construction of such a research base. More specifically, the aim is to serve these ends: First, to provide a sociological perspective on the problem of air pollution, drawing upon the general framework and the basic concepts of the discipline. Second, to call attention to those few studies that constitute the beginnings of sociological data on the subject of air pollution. And, third, to speculate, on the basis of the first and second points above, on the kind of research that would be most appropriate in the future.

ISSUES AND ANALYSIS

The Heritage of Human Ecology

Most biological scientists and lay persons who have recently turned their attention in such large numbers to problems of ecology and the environment would likely be surprised to learn that the field of human ecology has been one of the most important and enduring research areas of American sociology since the late 1920's. Unfortunately, however, this surprise would give way to disappointment when they learned that the ecological studies of sociologists have emerged from a framework which has made minimal use of the physical environment as an analytic variable.

The human ecology of sociologists can be traced to the theoretical work of Park (1925), Quinn (1939), and, more recently, Hawley (1950). The basic premise is that of a shared, limited, human habitat which generates a process of "competitive cooperation" in which the most strategic portions of the urban landscape devolve to those "dominant" land users who can most intensively exploit a given land parcel. This mode of analysis was built upon to generate explanations of land-value patterns (Hoyt, 1933; Schmid, 1940), and every conceivable social and psychological pathology of urban life (Burgess, 1925; Dunham, 1937). The ecologists believed that cultural and psychological phenomena could best be understood through study of the struggle over the use of land which, as they understood it, was determinant of other aspects of social life.

In his summary of the limitations of the ecological work, Michelson (1970:17), pointed to the following fact:

> Space has been utilized as a medium in most of human ecology rather than as a variable with a potential effect of its own. The environment has been seen as a flat plane, with occasional internal boundaries such as railroads or parks which could set apart natural areas, within which subsocially determined aggregates jockey indefinitely for turf.
>
> When speaking of environment, the human ecologists have referred to social environment. By what social groups or activities is a particular territorial aggregate surrounded? What difference do these social surroundings make for an aggregate's internal social structure and hence pathologies?

As Park (1952:165), himself explained it:

> Human ecology, as the sociologists would like to use the term, is . . . not man, but the community; not man's relation to the earth which he inhabits, but his relations to other men, that concerns us most.

Thus, in addition to all its other omissions (cf., Firey, 1945; Molotch, 1967), the human ecology school has failed to consider the physical environment as part of the system in which humans exist. It has failed to take the advice of Duncan (1961), that sociologists think in terms of ecosystems rather than social systems. Michelson calls attention to the problem which thus exists among the disciplines by citing a remark by Barker (1963:26):

Except in their applied phases, the biological and physical sciences have eschewed ecological units with human behavior as component elements. They have stopped with man-free ponds, glaciers, and lightning flashes; they have left farms, ski-jumps, and passenger trains to others. And psychology and sociology have, for the most part, shied in the other direction: They have avoided whole, unfractionated ecological units with physical objects as well as people and behavior as component parts. So behavior-setting-type units have almost completely fallen between the biophysical and the behavioral sciences, and this has been a source of serious trouble for the eco-behavioral problem.

Even when the biologists have attempted to bridge the gap by bringing people into their formulations, they tend, according to Duncan (1961: 144), to do the following:

[To] deplore and to exhort, not to analyze and explain. The shibboleths include such phrasings as "disruption," "tampering," "interference," "damage," and blunder," applied to the transformations of ecosystems wrought by human activities. . . . <u>They evidently need the help of social scientists in order to make intelligible those human behaviors that seem from an Olympian vantage point to be merely irrational and shortsighted</u>. [Emphasis added.]

The Theory of Cultural Lag

A move toward the understanding of the human ecosystem has been made by Duncan (1959), in his suggestive proposal of four categories as constitutive of the "ecological complex" of a human ecology: Population (P), Organization (O), Environment (E), and Technology (T). Each of these factors is interdependent with every other; the implication is thus that pathologies in the habitat can be understood by locating the casual dynamic operating among them at a given point in time. In this connection, the work of an important sociologist (not an ecologist), W. F. Ogburn, may be profitably examined.

Although he wrote several generations ago, Ogburn (1922) made a contribution to the conventional wisdom which remains wisdom, nonetheless. "Cultural lag," according to Ogburn (1957), occurs, "when one of two parts of culture which are correlated changes before or in greater degree than the other part does, thereby causing less adjustment between the two parts than existed previously."

By the illustrations he cited, Ogburn made clear, that the problem more specifically involves the frequent inability of innovations in social organization to keep pace with changes in technology. Ogburn was a "materialist" in that he considered the material or technological base of society to be the major, driving force of history. Technology was seen as having a life of its own--changing and moving--with the social system always in a race to keep pace with the newest innovations. Man's social organization is always in the process of adapting to realities newly created through technological change. The time it takes to catch up is the period of "cultural lag." If we adopt Duncan's schema, then Ogburn's "materialism" can be represented as $T \rightarrow O$, with the ($\rightarrow$) requiring lag time. Duncan (1961:148), stated:

> [We are] aware of the many complications concealed by the use of arrows linking the broad and heterogeneous categories of the ecological complex. The arrows are meant only to suggest the existence of problems for research concerning the mechanisms of cause, influence, or response at work in the situation so sketchily portrayed.

Whereas cultural lag may ordinarily lead to inconveniences and disutilities of various sorts under typical world-historical conditions, events of the post World War era make cultural lag a very dangerous phenomenon indeed. Ogburn feared that the invention of atomic weapons without major innovations in the social sphere might well mean the end of all life. The continuation of a system of armed nationstates operating without a ban on nuclear warfare (a self-constraint), was the social reality inconsistent with the technological breakthroughs of nuclear energy. Again, in Duncan's notational format: $T \rightarrow O$ or $T \rightarrow E \rightarrow P$. That is, either technology leads to change in social organization or, failing that, technology changes the environment in such a fashion that the population is destroyed, in this case, by fire, fallout, and so forth.

The ultimate environmental pollution would be the result of a nuclear war. And although we may be satisfied that the probabilities of such an event are low, the issue which that prospect raised for Ogburn remains one of the central problems in discussing any form of pollution. That issue is the proper relationship between technological change and social organization. Ogburn's writings imply that the solution is always one of "adaptation," i.e., adaptation of the latter to the "realities" imposed by the former (i.e., $T \rightarrow O$). Adaptation, of course, is an ambiguous word: Does it mean that men passively adopt a new kind of life style or personal habit because a new, more "efficient" invention has come upon the scene? Do we, for example, end outdoor recreation as a proper adaptation to the increase in the number of people driving cars and thus polluting the atmosphere ($P \rightarrow T \rightarrow E \rightarrow O$)?

Or might adaptation imply something else? Might it imply the development of social institutions that create control over new technologies so that technological gains are made useful in the service of socially defined ends (P→ T→ O → T→ E → P)? It is this latter meaning of adaptation that Ogburn's later writings seem to imply; it is this latter meaning that is taken to be the essential sociological problem facing those who wish to end air pollution: How can people generate social institutions that can make them the masters rather than the servants of their own technology?

Except for Ogburn's early contribution, sociologists have not directed their research efforts toward this problem. Indeed, the thrust of most social science research on technology is in the opposite direction: How can we gain behavioral adaptation to technological changes that are taken as given? Much of the anthropological literature on preliterate or semiliterate societies, for example, concerns itself with the "problem" of peasants and tribesmen who have difficulty "adjusting" to the industrialization that comes their way. The policy thrust of such studies is not in terms of discovering how the indigenous population might gain control of these events and shape them to their own wishes and cultural styles (O → T), but, rather, how the natives can be induced to make those decisions that will make the transition to industrial life (considered as necessary and inevitable), as painless as possible (T → O). Rural sociologists similarly tend to assume the "good" and the "necessary" of technological change; the problem is seen as simply gaining behavioral adaptation most quickly (e.g., Loomis, 1957; Rogers, 1960; Sussman, 1959). A new farm technique exists; the "problem" is how to get the rural population to make the "necessary" behavioral and social, organizational modifications. There is an analogous urban literature directed at the problem of gaining from urbanites those behaviors that are assumed to be required for successful adaptation to the exigencies of urban, industrial life. Thus, proper handling of personal refuse in an urban context becomes a problem in the cities of underdeveloped societies (cf., Clinard, 1966); in developed societies, the need is to change those subcultures at variance with modern technological society (cf., Silberman, 1964; Moynihan, 1965).

The Questioning of Accumulation

Regardless of the wisdom of this research approach when confronting these other matters, the constancy of the appraoch has perhaps blinded sociologists to the possibility that simple technological change--any kind of technological change introduced as "growth" or "advance" by its sponsors--may not result in the "ultimate good." Small children are taught in the school system that inventions are major triumphs of genius which appropriately force into being new kinds of social organization. Sociologists were also children, and it is understandable that they should grow into the typical adulthood of uncritical acceptance of increased "productivity." But productivity, even favorably defined as increasing levels of useful energies, is an extremely crude form of a societal goal. Ray (1968:18), pointed out the following:

> Stocks of energy do not control, though we often think of technology as giving us power (and therefore control) over nature. Rather, we are all familiar with the disruptions that ensue, being magnified through an ecosystem. . . . The failure of societies is in the realm of control, over the environment and over the mutually inconsistent behaviors within society.

In other words, man needs first to define his goals in terms of the pleasures he seeks and the qualities of life he desires, and then he needs to utilize technology so that there is "low utilization of energy to produce effects of major magnitudes within precision" (Ray, 1968:18). Technology needs to be the rational harnessing of energy to achieve with maximum efficiency certain life goals which are determined outside the context of technology and production. If a blue sky or sweet air is an ultimate goal, there is no reasonable argument that could dismiss such a goal as "impractical." It is practical in that it is primary. Given the creation of a productive base that is capable of maintaining basic subsistence, the primary, historical rationale for determining the use of resources is at its end. Ray (1968:19), has put the matter succinctly, and, in so doing, he provides a perspective which could form the basis of the sociological mission:

> If scarcity no longer is the critical problem, then sustenance technology must leave the driver's seat, and so must those organizations controlling it. . . . We are not prepared for intellectual paradigms that take affluence for granted and state the desirable (rather than necessary) as highest goals. . . . Futurism of a scientific kind can spell out the possible and the probable in alternative futures some day; but more is needed: utopian futurism of a quasi-novelistic kind, replete with imagery, but based on solid calculation.

What People Want

There is some evidence to suggest that the U.S. public is increasingly turning to life visions which challenge the goal of crude-energy maximization. The consumption of consumer goods as the

AIR POLLUTION AND SOCIOLOGICAL RESEARCH   21

basis for social organization has been challenged occasionally over the last several decades (see Galbraith, 1958, for one important version), but only recently have there been signs that massive numbers of people in the United States may be turning to a broader definition of the good life. There does exist an ever-increasing body of popular literature that documents this growing sensitivity by the public to societal goals located outside the private-consumption framework. Foremost among that evidence has been the sharp rise in public indignation and protest over the pollution of the environment. Pollution has at last arrived on the "most important problem" list of the Gallup Poll (Gallup Opinion Index, 1970a). Most striking is the fact, as Gallup has reported, that 73 percent of the U.S. people report a willingness to pay higher taxes in order to "fight" conservation problems" (Anonymous, 1969a). Similarly, a majority feels that "more tax dollars should go for natural resource programs... less for space, defense, and foreign aid" (Anonymous, 1970a). The quality of the environment has become a concern of leading business men (Ananymous, 1969b), as well as a major issue for college students (Gallup Opinion Index, 1970b; Anonymous, 1970b).

The newly invigorated conservation movement, like the militancy of youth, came as a surprise to many observers of the U.S. scene. For example, as recently as 1967, Loveridge (1969) held 45 interviews with a random sample of San Bernardino area residents on the subject of air pollution. His pessimistic conclusion was as follows:

> ...most people do not know how or even why they should become participants in public policy decisions on air pollution control.... [To] expect the public to take political leadership in the fight--barring of course an unprecedented disaster--is sheer foolishness.... Rather than waiting for a confused and preoccupied public to act, solutions will depend on effective leadership, coming from opinion and policy makers.

Loveridge could not have foreseen that pollution particularly air pollution--was to become a major political issue just one year later and that it would become one of the most important issue-bases for new community organizations across the country.

Additional studies have tested the public's reaction to air pollution. The problem with most of these researches is that they tend to rely on very small samples--sample sizes that prevent the kinds of analysis which would take one beyond a mere reporting of column-marginal responses to survey questions. In addition, these studies are not based on national samples. Rather, they are based on areas of the country in which the researcher in question happened to live or found convenient for other reasons. We are thus left in the position of having to compare findings gathered in Nashville, with those gathered in Scandinavia, with those gathered in Los Angeles. But variations in methodologies preclude seizing on geographic variations as an opportunity for comparative, secondary analysis. For example, the studies ask their questions in different ways: One cannot equate the phrases "are you bothered?" and "is it a hazard?" Such slight variations in wording may have significant impacts on responses; thus, without standardized questions, interstudy comparability is lost. In addition, the study locales suffer from varying extents of pollution (often unspecified) and differing types of pollution (e.g., smog, foul ordors). Finally, the studies published to date are out of date; none were carried out subsequent to the "boom" in conservation and air-pollution interest.

Nevertheless, for what they may be worth, indications of some of the study results follow: Smith, et al. (1964), reported that "up to" 3.8 percent of their Nashville sample voluntarily expressed awareness and concern about air pollution as a health problem, whereas 23 percent of the respondents stated they were bothered in some way by air pollution. It is likely that where there is extensive air pollution, the amount of "complaining" increases dramatically. Thus, in a 1956 study conducted by the California Health Survey, 66 percent of Los Angeles County residents reported that air pollution did "bother" them. In addition, 17 percent of Los Angeles County residents considered changing their residence because of air pollution, and 9 percent considered changing their job for that reason. It would seem reasonable to suppose, given the increase in attention paid to smog and the greater extent of some forms of it in the present day, that the 1956 figures, perhaps a "high" for that decade, understate the degree of contemporary dissatisfaction. An additional series of studies conducted over the 1960's tends to indicate, regardless of whether the study locale was St. Louis, or Clarkston, Washington, that an increasing amount of complaining is taking place. (See Crowe, 1968; de Groot, 1966 and 1967; Jonsson, 1963; Medalia, 1964 and 1965; Rankin, 1969; Schusky, 1966; Stalker, 1967; U.S. Dept. of HEW, 1969; Williams, 1966.)

The best of the air-pollution surveys, and also the most recent to date, was carried out by Gold (1970). Based on a random sample of Californians, Gold's results confirm the growing dissatisfaction with smog and specify the salience of the problem, the wide range of daily activities it is seen to influence, and some of the ends to which Californians would go in order to ameliorate the problem. Respondents were asked: "Here is a card listing some problems that people often talk about. I would like you to look through these and select the three you think are most serious these days." Gold found that 59 percent selected air pollution as among the three most serious problems.

This was only 4 percent below that proportion which listed "crime and violence in the streets" as among the three most serious problems. Among the higher-income groups, the young, the higher-educated, and those living in Los Angeles and Orange Counties, air pollution was the problem most likely to be chosen as among the most serious. In terms of their routine activities, significant proportions of Gold's respondents reported that smog had effected their choice of workplace, shopping habits, frequency of washing their car, the upkeep of house and housecleaning, and spare-time outdoor activities. It is thus not surprising to learn that overwhelming majorities of Gold's respondents favored curative action including (but not limited to) diversion of some state gas money for air-pollution research and required annual inspections for all cars.

There is thus good evidence to suppose, in the case of air pollution, that the voiced concern of respondents in sample surveys bespeaks a definite, felt trouble and that it is not a vague and amorphous issue detached from personal experience. People not only think that their daily routines have been altered by smog; they also think that smog is a health danger. Gold found that 69 percent of his respondents believed that "smog has reached a point where it is a danger for normal, healthy people," and the proportions who believe this are still higher among those in high-smog counties and those with higher levels of education.

Although air pollution may damage the life of everyone living in smog areas, it becomes a high-priority problem for those people who have solved most of their other problems. It is one of those problems that arises from the need to cognize goals in terms other than "survival" through consumership and basic health care--a need which falls only on those persons who have satisfied the more basic needs. Gold reported a positive and direct relationship between income and propensity to list smog as a serious problem. The poor were almost as likely as the more well to do to believe that smog is a health danger, but, unlike the rich, they found other dangers (e.g., unemployment) to be of relatively high significance. Medalia (1964) found that professionals, managers, and proprietors were far more likely to be aware of pollution and to consider it a serious problem than were respondents in the lower-occupational categories. Similar findings were reported by Smith, et al. (1964) and Devall, (1970). Van Arsdol (1964), reported that blacks in his Los Angeles sample were less likely than whites to report being bothered by smog.

When we speak of air pollution as a social problem, we are thus speaking of a particular (although very large) social group for which the problem is paramount. It is among those of the white race, with higher incomes and more education and among the youth that salience of pollution as a problem is highest. These are precisely the individuals who tend to be the most active politically and organizationally. Medalia (1965) learned that the more respondents were involved in or identified with their community, the more upset they were about its polluted air. Similar findings were reported by Van Arsdol (1964). Thus, the implications are that the constituency for public action on air pollution comes precisely from those groups and types of persons who are most prone to take problem-solving actions in the public arena. How well traditional, political activities will fare in the context of pollution battles is a major question and one to which we now turn our attention.

Clean Air as a Social Movement

Thus, we arrive at the tentative conclusion that among those in the United States who have lived with their basic needs of food, clothing, and shelter satisfied, there is emerging a new consciousness of alternative goals. These goals would certainly be difficult to state in their entirety, but among them would be a physical environment which is healthful and aesthetically pleasing. Medalia traced the source of this new kind of personal goal.

> The individualism and sensual repression characteristic of Victorian society contributed to that society's high capacity for material production, both of capital goods and of air pollution, and they also may have contributed to a low awareness of air pollution as a social problem. . . . Today, broad changes in the social structure and ideology of American society have given rise to a generally increased awareness and a generally lowered tolerance of air pollution. . . . industrial pollution may no longer be regarded as positive evidence of success in production, so much as evidence of lack of success in consumption, or "good living"   [Medalia, 1964:165].

And although there have not been detailed studies that attempt to examine pollution and youth activism, the growing concern over the environment, if not all its origins, may well be part of the same historical process which has generated the student movement. Flacks (1967), and others clearly demonstrated that the ranks of the student activists had been filled from the same social class as to respond most negatively to pollution--the upper-middle class. Flacks' findings indicate that the politicization of youth was anchored in the child-rearing practices which stressed creativity and experiental accomplishments rather than the neo-Puritan, hard work, consumption ethic. With only Gold's data to confirm that youth is disproportionately aggrieved over polluted air, we can only speculate on the relationship between the ecology movements and youth activism in general. Research in this area would clearly be of some importance.

In his case study of the Santa Barbara oil-pollution disaster, Molotch (1970), attempted to trace out the mechanisms by which environmental concern was converted to environmental activism. Although the membership of the Santa Barbara activist group was in some dimensions different from that of other militant groups (such as blacks or students), the white, upper-middle class and upper class Santa Barbarans found themselves recapitulating the historical, tactical development of those other movements as they fought (unsuccessfully) to rid the Santa Barbara Channel of oil drilling. They found themselves going from organization and community meetings, to petitions, to protest, to picketing, to demonstrations--and finally to civil disobedience. The rather extraordinary militancy of the Santa Barbara conservationists may make sense in light of the fact that in Santa Barbara, the new, emerging life style outlined above is, in some sense, the dominant life style: It is a community of people who have chosen the area precisely because of its environmental quality. Thus, the great oil spill represented a sudden deterioration in environmental quality for persons whose lives were built around the pleasantness of their environment and who well remembered the days when Santa Barbara was completely unscathed by any oil activity. And this is why the militance of the Santa Barbarans may be a harbinger of the kind of activism that is to come when future confrontations appear between persons who value environment and events that result in environmental despoilation.

Medalia (1964: 162), found, even in a town where there was no dramatic event (such as an oil blow out), but where pollution was a more routine matter of fouled air from a pulp mill, that there still existed a "relatively high potential for situation-altering actions" relative to the source of air pollution. That is, Medalia's respondents who thought that pollution existed also were likely to think that the level of pollution could be reduced, and nearly three fourths of these respondents thought that the firm polluting the air was not doing all it could to end the problem. Even so, the degree of militance does not nearly equal that of the Santa Barbara citizens (who were virtually unanimous in their demand that oil companies be forced out). In the town studied by Medalia, only about a third of the respondents would go so far as to say it would be appropriate to "ask their elected officials for effective controls on air pollution." As might be anticipated, those most concerned about pollution were most in favor of this "strong" action, whereas others tended to favor the curative mechanisms of "support of industry efforts" and/or developing "more information" about the pollution problem.

Again, it is important to note that Medalia's study as well as the Santa Barbara incident occurred prior to the great wave of publicity and organizing about the pollution problem. There is thus a need to develop contemporary information paralleling that described above-- information regarding the demographic, socioeconomic, and geographical distinctions among those who respond to the environment in varying ways. We need analysis of the interrelationship of these variables as well as analysis of the pattern of interaction of such variables with the community contexts in which they appear. That is, we need to understand the relationship between these conventional variables and the extent and type of actual pollution, the perceived extent of pollution, and the social dynamics of the region in question. A series of case studies of movements and organizations now in motion on environmental issues in general and air pollution in particular should augment any additional surveys on this question. Such organizations exist all over the country and include, in addition to the more traditional groups such as The Sierra Club and The Audubon Society, such newly formed activists' groups as Ecology Action, Friends of the Earth, Get Oil Out in Santa Barbara, and Save the Bay in San Francisco. These organizations can be partially viewed as part of a social movement that is attempting to reorder U.S. priorities and provide a vision of alternative life styles and alternative ends for resource utilization. They will likely play an important role in future activities in environmental politics, including air pollution legislation, research, and standards enforcement.

Clean Air and Community Organization

The new organizations that are springing up in response to environmental pollution can be viewed as a possible vehicle for effecting the changes in law, industrial policymaking, and political organization which are necessary for a solution to air pollution problems. An extensive literature on the subject of mobilizing people toward specific goals through community organization has been developed by sociologists, social workers, and community-organization practitioners.

The academic literature on community organization springs from the early research of Kurt Lewin (1958), and his students (Lippitt, et al., 1958). Lewin's experiments attempted to find how best to induce people to adopt new perspectives and habits so that socially desired ends might follow. In the context of the air pollution problem, those ends are twofold: First, to generate organizations that can affect the political structure to promote pollution-control measures, and, second, to generate everyday behavior habits that are compatible with a clean-air environment.

Lewin found that people were most apt to change their ways if they conceived of the new decision as emerging from a group of their peers. Lewin and his students organized a now famous series of experiments aimed at determining the social conditions under which decision to change could most easily be effected. In one experiment, aimed at convincing women to serve intestines to their families (as part of the war effort), group decisionmaking was compared with the lecture method. With the help of an expert to "guide" the discussion, women who examined the problem "for themselves" in informal discussion with other women, were found to be far more likely to actually serve intestines than were those women who were lectured on the nutritious and patriotic advantages of the same substances. Similarly, a Lewin experiment aimed at changing new mothers' feeding habits documented the superiority of group decisionmaking over one-to-one counseling.

The theory of community organization thus recommends the creation, as an on-going enterprise, of the kind of group dynamic for social change that took place in Lewin's experiments. Individuals are seen to be most likely to change when they have the opportunity to discuss "the problem" with their peers and when the decision for change develops logically and deliberately from considerations which they themselves have raised. Insofar as the focus of action is internal rather than external, the method is termed self-help, and it has been tried in the United States and around the world, particularly in low-income areas. The results appear to be mixed, and although there have been extensive evaluation studies completed on community-organization projects (see Clinard, 1966), there still remains ambiguity as to whether or not the traditional community-organization techniques are efficacious. There are very few such organizations that continue after the organizers leave the scene.

An alternative model was proposed by Saul Alinsky (1946), founder of the Industrial Areas Foundation. Alinsky's basic concept was that before there could be a feeling of community, of "we," there must be an out group, a "they." His strategy was thus to focus on a hostile agent and to rally support around a new organization which was seen as the collective vehicle for destroying (or at least neutralizing) the enemy. Alinsky had "successes." In the late 1930's he was able to organize a working-class, white community in Chicago into an effective grass roots organization know as the Back-of-the-Yards Association. This group was the spearhead for the physical rejuvenation of the community, for an increase in levels of public service, and for the exclusion of minorities from the area. A second Alinsky success, created in late 1960, was the all-black Woodlawn Organization (TWO), which was able to thwart expansion plans of the nearby University of

Chicago. Both organizations, although transformed and modified over the years, remain active in Chicago affairs.

Both the Alinsky model as well as the more traditional models of community organization are available for those fighting air pollution and similar environmental ills. Both types of organization seem to be developing within the environment movement--sometimes within the same organization. Thus, conservation groups develop political goals of affecting an external enemy (e.g., government and/or industry, as in the Santa Barbara case), as well as more traditional techniques of self-help (e.g., trash recycling and bicycling).

It may be that the U.S. tradition of "voluntarism" can be harnessed to solve pollution problems. Community-improvement associations and neighborhood-protective societies might be able to restructure themselves into environmental police forces, capable of bringing official attention to polluters or as coordinators of other environmental-protection programs. Similarly, such groups could serve as watchdogs over public agencies charged with air pollution responsibilities. The research need is to learn which kinds of community-organization approach will work most efficiently for the publics that form the constituency for the environmental movement. Studies based on ethnographic accounts of existing organizations might attempt to pinpoint bases for success and failure. The entire literature on community organization could be made relevant to the air-pollution issue by investigating current attempts being made to alter or reverse environmental policies. Mechanisms need to be devised to channel technical information in formats that are useful for these organizations and that will enable them to function with maximum rationality, based on information from authoritative sources.

Air Pollution and the Structure of Power

The attention which has been given to social movements and community organization in the discussion above is rooted in the authors' assumption that ending the problem of air pollution is not simply a matter of gaining the right kind of information and automatically effecting policy change based on that information. Rather, the assumption is that there are vested interests in the continuation of air pollution, and that the problem of pollution is at least partially, if not preponderantly, due to that fact. The controls of resource exploitation and production in the United States (and much of the world) are moved by a (relatively) few persons. And, because discussions of ending air pollution usually imply a rather massive change in the methods of resource exploitation and production, the fate of the air is in the hands of what Ray (1968: 17), has termed, "a small number of decisionmakers who happen to be among the most powerful men in the world." Ray asks a very disturbing question:

> Who is powerful enough to "control technology" when it means restricting giant corporations and major government agencies--other than the same men in the power structure? <u>Et quis custodiet ipsos</u>?

[He answers:] Control had best be called something else, or nothing will be done. It may as well be called "planning" [Ray, 1968: 17].

Unfortunately, there is much in the sociology-of-planning literature to suggest that it may be called "planning" and still nothing will be done. There is always the danger that planning will serve as a smokescreen of expertise to cover up the political process by which the most powerful bargain among themselves for maximum profit and advantage (see Meyerson, 1955; Banfield, 1961). Rather than simply calling it planning, the hope for control over the controllers rests on the same bases as any other hope for making democratic procedures real. Somehow the weight of an informed and mobilized public must become supreme. To change the name to "planning" may or may not prove a tactical advantage, but in itself it could only turn out to have trivial consequences.

How difficult is the task of asserting power over the powerful in relation to air pollution? The answer turns on several issues, including, of course, the magnitude of the financial costs involved and the extent of the sacrifices which various powerful groups will come to experience should any relevant change be effected. Research should provide an analysis of the total costs of achieving various levels of air purity as well as an analysis of who will pay those costs. The sociologist must learn which sectors of the industrial world will tend to lose through efforts to clean the air and which will gain. Important are not only the actual gains and losses which are likely, but the gains and losses as perceived by the decisionmakers themselves. The payoff of such a research activity could be one or more of the following: First, those individuals who exaggerate the costs to themselves could be informed of the realities of lower costs. Second, for those who would pay the high costs, efforts could be made to achieve alternative strategies that would provide least-cost consistent with clean air; we have in mind here research analogous to studies for reconversion (or simple conversion) from wartime to peacetime industrial activity. And third, potential supporters of clean air could be located and brought into mutual contact.

It should be noted that these strategic formulations rest on the rather benign view of U. S. power held by some (although not all) sociologists and political scientists that U.S. power is basically pluralistic. That is, power is not located in the hands of a single, cohesive, economically dominant elite. Rather, the argument goes, power is located in the hands of a series of subelites who collectively wield power, but who, at least on some issues, come to have disagreements that lead to competition for power. Summary reviews of this assumption and its implications are contained in Merelman (1968), and Mankoff (1970). Without a detailed analysis of the pluralism arguments pro and con, let us direct attention toward the implications for pollution control were it true that pluralism existed. The strategy that logically follows would involve finding out if any of the subelites which have some power perceive clean air as advantageous to their interests, and, second, forging organizational mechanisms and political strategies to provide such a clean-air coalition with a basis for effective participation in the political arena. If, on the other hand, we were to make the assumption that pluralism does not exist and that the United States is ruled by a power elite (Mills, 1956), the strategy must either be to convince that ruling elite that air pollution is bad for them, to appeal to their sense of charity, or to modify the system which enables them to rule and ignore the people's desire for clean air.

Air Pollution and the Mobilization of Bias

In order to create a victory for clean air, regardless of the power realities that provide the context for the battle, there are certain strategic clues which can be brought to light and, through further investigation, perhaps eventually be refined. E. E. Schattschneider (1960), provided the basis for sensitivity to the fact that the first step in achieving a political end must be to have one's position become part of the battlefield. That is, there can be no success in the political arena if one's problem never even gets raised; there are many possible issues which are never mentioned in a given political system. For a problem to become a major issue, a vast number of persons--newspaper editors, congressmen, television commentators, and opinion leaders in industry and the academy--must turn their attention to the same matter. Only a few such issues can be dealt with simultaneously. There is thus the question of the processes that determine what becomes an issue and what does not.

> All forms of political organization have a bias in favor of the exploitation of some kinds of conflict and the suppression of others because <u>organization is the mobilization of bias</u> (emphasis added). Some issues are organized into politics while others are organized out [Schattschneider, 1960:71].

The question clearly suggests itself in the case of air pollution. How could the issue itself lie dormant for so many years, especially in such a smog-ridden setting as Southern California? How is it, "out of the blue," that this issue of environment arises to become one of the major bases of many political campaigns in the United States? A careful study is needed of the process by which this occured. We need content analyses of the major media to discover how the issue arose and grew, the contexts in which it was discussed, and the responses by "readers" (in letters and columns, for example). We need to chart the growth of coverage, the relationship between that coverage and media comment, and the adoption of pollution as a "talking point" in political campaigns. Air pollution, as Loveridge (1969), has indicated, was once not an issue; it is conceivable that it could become a nonissue once again. The process of the dynamics of issuemaking should be studied in hopes of avoiding that fate.

The making of an issue is not, by any means, synonymous with effective action to satisfactorily settle the issue. Edelman (1964), pointed to what he termed the "symbolic uses of politics": What is discussed in public arenas as the basic "issue" may provide merely the symbolic talk that operates separately from the real decisions producing and influencing power. The "real" decisions and the "real" issues may have to do with the struggle among competing elites for the governmental rules and decisions that they need to carry on their work of effective and profitable resource exploitation and goods production. In this perspective, disputes over pollution, although authentic in the public arena, are, like controversies over Communism and law and order, mere sideshows to the real process of governmental decisionsmaking. No doubt real consequences flow from these symbolic battles (men get fired, wars may even start), and real tempers become quite short. But the important business of government--the decisions that determine fiscal policy, that allocate resources for exploitation, that effect the distribution of the tax load--these decisions occur in relative quiet, among those who are real participants in the bargaining for maximum advantage. Edelman considered the outcomes of the symbolic battles often to be relatively trival (e.g., civil-rights laws which are unenforced; legal, procedural guarantees which have no real impact, certainly no real impact on the functioning of the U.S. political economy). It may or may not be that the "right" decisions are ultimately made; Edelman's point is that this decisionmaking process bears little relationship to the "issues" which are part of the public controversies.

It would seem that the phenomenon of air pollution possesses attributes which make it an ideal case study for investigating some of the matters raised above. Certainly an examination of decisionmaking about air pollution would help to provide the connection between actual policymaking and administration, on the one hand, and public controversy, on the other. A tightly drawn research investigation of the history of the air pollution controversy, its subsequent legislative disposition, and its administrative consequences would be a major contribution both to the literature on air pollution and to the social sciences as well.

The Administration of Solutions

Even in the event of major breakthroughs--either in the area of technology or new legislation--solution of air pollution problems would consist of the long-term functioning of institutions devised for the purpose either of setting minimum air-quality standards, affixing penalties (either market costs or criminal sanctions), or recommending and enacting the kinds of laws which subsequent changes in technology or society might require. The need is thus to build the sorts of institutions that will be able to function effectively over time and to adapt to new conditions as they arise.

An activity such as the establishment of minimum air-quality standards is itself a social process and one that has important implications for the success of environmental improvement efforts. Shiffman (1967) carried out a study of some pollution-control programs (not on air pollution, per se), which, in terms of stated research goals, can serve as a model for the types of investigations that are needed in the analysis of standard-making processes. His topics of investigation are as follows:

1. The factors in the development and application of standards that are significant in the administration and operation of environmental pollution-control programs. The focus is on the political, social, and economic determinants of the standards process and on the the role of administrative behavior in the use of standards.

2. The conceptions of and attitudes toward standards which are held by the technical specialists and administrators who are responsible for the implementation of the standards in the operation of environmental control programs.

3. The influence and relation of these conceptions and attitudes on the purposes and manner in which the standards are employed in program operation [Shiffman, 1967:8-9].

His conclusion is not a very happy one but is most heuristic for the social scientist: "It seems painfully apparent that the establishment of standards (as presently done) is neither a rational nor consistent undertaking" (Shiffman, 1967:52).

More generally, there exists a social-science literature on the internal and external functioning of those regulatory agencies that have thus far been brought into existence. The classic work was the examination by Selznick (1949), of the Tennessee Valley Authority. It has been concluded that regulatory agencies, given their internal "need" to grow and protect themselves from adversaries, tend to become the guardians of the very industries they were designed to regulate. The problem goes beyond that of selling out or corruption; it is a matter of well-meaning individuals who make the kinds of compromises that they consider necessary to maintain and enhance the organizations which they serve. Organizational goals (as formally instituted), are thus displaced by other goals that have to do with organizational survival and personal career advancement (cf., Sills, 1957).

These general difficulties entail a series of questions that could be answered through a program of air pollution research:

1. How can institutions be structured so that members and staffs of regulatory agencies can advance in their careers without having to defer to the regulated? Can there be professional careers contained entirely within the regulatory apparatus without the assistance of regulated industries and agencies?

2. How can those who serve on regulatory boards be insulated from the influence (explicit and implicit) of those who are regulated?

3. What are the consequences of the presence of industrial representatives on control boards and commissions?

4. Can experts be found who are not associated with those industries to be regulated and whose implicit assumptions on the nature of resource management differ from those of the regulated?

5. How successful has voluntary compliance been as a method of air-quality maintenance? What modifications and/or alternatives are available?

Pollution solving is inevitably going to be a process, not an act or series of acts accomplished once and for all. Any programs that are instituted will have to emerge from a recognition of the process nature of the problem. Sociologists may be able to draw from the rather extensive literature on what is termed "formal organization" in order to provide insights into how maximum effectiveness might result from various organizational arrangements. Concomitantly, studies of existing regulatory organizations in this country, as well as those abroad, should be carried out in a comprehensive search for the ingredients of successful "solution-doing" institutions.

TASKS RECOMMENDED FOR AIR POLLUTION RESEARCH

Survey Analysis. What would well-conducted surveys, carried out subsequent to the "boom" in popular interest, indicate about the demographic, personality, socioeconomic, geographic, and community characteristics of air pollution "complainers?" What are the differential responses of social groups to different components of air pollution? What is the level of awareness concerning air pollution, its various forms and components, and the concrete dangers to health and the economy that each form of pollution entails?

Community Organization. What are the missions, accomplishments, and methods of existing, voluntary conservation (particularly air-quality oriented) organizations? Which forms of community organization work most effectively? How can mechanisms be devised to channel technical information in formats that such groups can use?

Interest Analysis. What are the interests that different social and economic groups have in given environmental policies? Exactly how do actions and policies that result in irrational outcomes for the community as a whole emerge from individual decisions that are rational in their original context? What potentials are there for coalitions among those who do, in fact, find their individual-rational interests served by clean-air policies?

Mobilization of Bias. How did air pollution as a public issue come into being? What is the relationship between issue and action? What would content analysis of major media indicate to be the dynamics of issue making and the prospects of the air pollution issue in the future?

Decisionmaking. How are decisions that affect the environment made? What would interviews with important ecological decisionmakers indicate about the geographical and social structural loci of such decisions? What is the social process by which such decisions are made?

Public Policy making. What are the social and political impediments to new environmental policies? How does decisionmaking in the private arena, as well as city, state, and national government, function to inhibit or facilitate new forms of environmental intervention?

Comparative Study Nonindustrial Societies. How have previous social systems dealt with their environment? What do historical and anthropological materials indicate to have been the determinants of given orientations toward the physical world both immediate and distant? What have been the differential consequences on rational and humane use of the environment of such variations in perspective?

Comparative Study of Industrial Societies. What has been the experience of other industrial nations in handling problems of the environment? What are the features of the U.S. political, economic, social, or cultural system that would likely to induce or inhibit the adoption of successful foreign approaches?

The Sociology of Standards. What are the social processes that determine the establishment and administration of air-quality standards? How do personnel within control agencies view such standards, and how do such views influence job performance and air-quality outcomes?

Career Opportunity and Socialization. What are the career routes that lead to environmental, regulatory tasks? Are such career routes structured to induce normative understandings that are consistent with clean-air objectives? Do viable career lines exist within air-control agencies so that professional personnel are not encumbered by career obligations to those who are regulated?

Institutional Insulation. How have pollution-control agencies succeeded or failed in their ability to insulate themselves from industries and agencies they were designed to regulate? How can institutions be designed so that this insulation is maximized?

Institutional Innovation. Can air pollution control agencies be structured to guarantee flexible responses to changes in the nature and extent of air pollution in a given locality? What do past organizational experiences indicate to be the ingredients of successful attempts to institutionalize innovation?

Enforcement Techniques. What has been the experience of agencies that have attempted to use voluntary compliance as a means of gaining conformance? What sorts of enforcement alternatives are available, and what have been their consequences when tried?

Visions. What are the kinds of clean-air alternatives, entailing more massive forms of social reorganization, that an all-out commitment to quality environment might involve? That is, what possible visions of new forms of production and distribution arrangements could be offered in the form of utopian ideal-types? What are the most efficient means toward the achievement of such ends?

CHAPTER

3

PSYCHOLOGICAL
AND BEHAVIORAL EFFECTS
OF AIR
POLLUTION

Robert W. Reynolds

INTRODUCTION

A survey of the literature on the psychological and behavioral effects of air pollution indicated a different pattern of effort in the two countries where most of the work seems to have been done. In the Soviet Union, research on maximum permissible concentrations of air pollutants was conducted on a highly structured basis and yielded a large body of data on which their standards were based. Because psychological data play a negligible role in the determination of standards in the United States, research efforts here have tended to be sporadic and not well coordinated. Because the Soviet data emphasize sensory thresholds, they yield standards that are probably unrealistically low. The U.S. standards, however, are probably too high, because they are based primarily on tissue pathology. It is recommended that a concerted effort be made to obtain reliable data on the effects of pollution on psychological and behavioral processes at a point where there is a deterioration of performance. In order to make generalizations to humans more reasonable, the first part of the effort should be devoted to long-term studies on the effects of various air pollutants on the behavior of monkeys. Subsequent studies should verify the applicability of the levels determined in the animal studies through short-term studies with human subjects.

ISSUES AND ANALYSIS

A survey of the literature since the early 1960's on the psychological and behavioral effects of air pollution reveals an interesting pattern of work devoted to this problem. If we restrict ourselves to the more careful inquiries and ignore the anecdotal, casual, and subjective reports, we find, with a few exceptions, that the work done on this aspect of the pollution problem in the United States and the western world has tended to be relatively fragmentary and incidental (Stokinger, 1969). On the other hand, research efforts in the Soviet Union have systematically examined the effects of various air pollutants, and combinations thereof, on conditioned reflexes and other manifestations of higher central nervous system functions (Detrie, 1969).

This difference in research effort on and concern with behavioral effects of pollution between the United States and the Soviet Union is a direct result of an explicit difference in the criteria used by the two countries in establishing standards for the maximum permissible concentrations of various air pollutants. Standards in the United States have been based primarily on levels of pollutants that induce pathological conditions, primarily in the pulmonary system. The Soviets, however, have emphasized levels of pollution that induce changes in behavior and electrophysiological responses in the central nervous system. Because the thresholds for such psychological and physiological changes are much lower than the thresholds for pathological changes, the Soviet standard is often one fifth, and occasionally, as in the case of ethylene oxide, as low as one ninetieth that of the U.S. standard (Levine, 1964; Truhaut, 1965). The political, economic, and sociological factors responsible for this disparity in standards would in themselves make an interesting subject for study and conjecture, but the important point here is the effect this disparity has had on research efforts.

Studies in the Soviet Union

Research on the effects of air pollution on behavior and psychological functioning in the Soviet Union has been conducted primarily under guidelines and procedures recommended by a permanent commission established by the Ministry of Health in 1949. The basis for the current criteria for air pollution control in the Soviet Union was discussed by V. A. Rjazanov (1965). Rjazanov described four possible criteria: First, concentrations of pollutants should not deviate from the levels found in a natural environment: This was rejected on the ground that such deviations may not necessarily have any adverse effect. Second, concentrations of pollutants would be acceptable that were below the levels which could be removed using current technology: This was rejected because it amounted to an endorsement of a technological status quo and provided no incentive for improvement. Third, the standard for air cleanliness should be that which is economically justifiable, i.e., it should cost no more to remove the pollutant than the cost of the damage caused by its presence: Such a

PSYCHOLOGICAL AND BEHAVIORAL EFFECTS

crass commercialism is rejected by Rjazanov as obviously incompatible with all humanitarian principles. Rjazanov's final criterion was the following:

> Only such concentrations of air pollutants as do not directly or indirectly exert a harmful or unpleasant effect on man, reduce his working capacity, or have a negative effect on him physically or mentally can be accepted as permissible. (p. 390)

According to Rjazanov, this criterion has formed the basis for air pollution control in the Soviet Union.

Rjazanov points out, however, that such a criterion must, in practice, be regarded as the ideal, because temporary and local conditions frequently result in unaviodable deviations from this goal. Nevertheless, it is on this criterion that the procedures for establishing standards were based. Because of this, the procedures used are very strongly oriented toward the determination of sensory thresholds and other psychological effects.

The maximum allowable concentration of any air pollutant or combination of pollutants is established by a set of procedures which are selected according to their utility for the particular pollutant in question. For example, the standard of 0.005 parts per million parts by volume for hydrogen sulfide is based on the human odor-perception threshold. This test, of course, would be useless in determining standards for odorless compounds, such as carbon monoxide, for which the threshold for inhibition of a conditioned reflex in animals provides the principal datum

There are two chief classes of procedure. The effects of single exposures to a pollutant are determined primarily on a series of tests on human subjects. These include thresholds for electrocortical reflexes, odor perception, respiratory response, dark adaptation, and chronaximetric response. In addition, data on clinical experience with animals are also used in setting single-exposure thresholds.

The data from single-exposure tests are normally supplemented by data from tests of the effects on animals of long-term exposure to the pollutants, often up to three months. Most of these tests seem to have been done on rats. The tests involved examination of changes in conditioned reflexes and electroencephalographic responses to stimulation, along with various tests of biochemical and histological changes. Here again, where possible, data from clinical observations of humans and animals are used to provide additional information.

The use of these procedures by the Soviets is predicated on the fact that changes in behavioral and psychological measures occur at levels considerably lower than those necessary to induce tissue pathology. Indeed, some of the tests, such as the human electrocortical reflex threshold, are used to establish effects of air pollutants at levels below that accessible to subjective report. This procedure involves the use of human subjects who show a strong alpha rhythm in the resting occipital electroencephalogram. The subject is tested in a darkened room. After breathing the gas mixture containing a controlled concentration of the pollutant for ten seconds, a light is flashed on for five seconds, blocking the alpha rhythm. Gas concentrations are used that initially do not by themselves produce alpha blocking. After about seven such trials, the alpha blocking begins to occur before the light comes on. A threshold is then determined to find the gas concentration below which such conditioned alpha blocking does not occur. Using this procedure, it was found that the olfactory threshold for sulfur dioxide is 1.6 mg/m^3 while the electrocortical reflex threshold is 0.6 mg/m^3 or 0.62 versus 0.23 ppm. We may also note here that the Soviet standard for sulfur dioxide is 0.26 ppm while the Los Angeles first-alert level is 3 ppm (Stern, 1964; Dunod, 1968).

Studies in the United States

As we have indicated above, because U.S. standards are based primarily on the induction of tissue pathology, there has been no organized and systematic evaluation of behavioral and psychological effects of air pollution in the United States comparable to that done in the Soviet Union. However, there have been research programs in various laboratories that have yielded important information.

Human Studies

The use of eye irritation as a measure of the effect of irradiated automobile exhaust has been studied by Hamming and MacPhee (1967), and Buchberg, et al. (1963). The latter group, in one of their studies, noted that odor thresholds may be lower than eye-irritation thresholds, and this factor may contaminate eye-threshold studies. Orcutt (1967), found that there was a linear relationship between log concentration of irradiated auto exhaust and probit percent incidence of eye irritation. Owens and Punte (1963), also showed that particle size was an important factor in eye-irritation effects and that large particles (60μ) had less effect than small (1μ) aerosols. Various other investigators studied the effects of several common air pollutants on other aspects of sensory, motor, and psychophysical functions. For example, Langerwerff (1963), used a battery of visual tests that included visual acuity, steropsis, lateral phoria, vertical phoria, divergence, convergence, visual field, and night vision in studying the effects of ozone. Weitzman, Kinney, and Luriator (1969), showed that chronic low-level exposure to carbon dioxide affected night vision and green-color sensitivity. Stopps and McLaughlin (1967), and Vernon and Ferguson (1969), studied the effects of tricholoroethylene on various tests of psychomotor performance. In a study of the possible interaction of temperature on odor threshold, Stone (1963), found no effect, presumably because the inhaled air was very rapidly equilibrated to body temperature in the nasal cavity. On a slightly more complex behavioral level, Beard and Wertheim (1967), showed a deteriorating effect of carbon monoxide on human time discrimination that was consistent with their observations of the effects on timing behavior in the rat as measured by performance on a differential reinforcement of long latency responses (DRL) schedule. This is a schedule of reinforcement that requires a pause of a finite period of time between responses in order to obtain the reward. Finally, Hore and Gibson (1968), found no effect of a 70 minute exposure to ozone (0.2 and 0.3 ppm) of performance on a standard intelligence test.

Animal Studies

The studies on animals have used rats and mice almost exclusively. Emik and Plata (1969), and Campbell, Emik, Clarke, and Plata (1970), observed pollutant effects on running-wheel activity in mice. Goldberg and Chappell (1967), observed the effects of carbon monoxide on continuous reinforcement (CRF), schedule in rats. Teichner (1967), showed a deteriorating effect of carbon monoxide on the behavior of rats in a runway, shuttlebox, and CRF schedule in a Skinner box. Bevilacqua and LaBelle (1963), showed that an excess of negative ions improved learning rates in rats while positive ions impeded learning. The addition of carbon particles produced improved learning rates over those produced by the ions alone. Beliles, et al. (1967), observed no effect of 0.08 mg/m^3 of mercury vapor on multiple FR-60 (60 responses per reinforcement), and FI-15 (15 seconds between reinforcements) schedules in pigeons. One of the more promising programs for the study of pollutant effects in animals appears to be that of Xintaras (1968), who was investigating the effects of carbon monoxide, ozone, and lead on behavioral, neurochemical, and neurophysiological (electrocortical) responses in monkeys. Unfortunately, the report available thus far was only descriptive of the proposed research program. Finally, Reynolds and Chaffee (1970), in a pilot study for the University of California's Project Clean Air, demonstrated a significant elevation of both simple and choice reaction time in monkeys after a half-hour preexposure to 0.5 ppm ozone.

General Discussion

Pollution is a relative term. It pertains to a continuum of effects ranging from the level where a single man, by the very act of breathing, is polluting his own atmosphere, to the point where an environment is rendered incapable of supporting human life by the accumulation of industrial, automotive, human, and other forms of waste products. Obviously, standards for acceptable levels of pollution must be set somewhere between these extremes. In view of the fact, however, that the imposition of any standards at all will necessarily involve a restriction of the freedom to act for various individuals and institutions in our society, the recommended standards must be based on a clear demonstration of the relationship between the level of pollution and its effects on man.

The Soviet standards, insofar as they are based on absolute sensory thresholds, are probably unrealistically low. For example, the point at which an auditory stimulus becomes distracting and has a disruptive effect on behavior is considerably above the absolute auditory threshold. In addition, however, this point is presumably well below the point at which the stimulus begins to produce tissue pathology. For this reason, if we may generalize from our knowledge of the effects of auditory stimuli to the possible effects of air pollutants, the U.S. standards may be set too high. Indeed, we must agree with Rjazanov's (1965) comment that tissue pathology data "cannot be of direct use in the establishment of maximum permissible concentrations of pollutants, as they tell us only about known harmful concentrations, and nothing as to how low a concentration must be in order to be harmless" (p. 396).

The use of human subjects in pollution studies is limited on the one hand by the dangers inherent in subjecting humans to unknown toxic effects in short-term studies, and, on the other hand, by the near impossibility of obtaining a significant number of subjects for long-term studies of several months' exposure to pollutants in a controlled, environmental situation. One approach to this problem has been to conduct these studies on animal subjects and to generalize the results to man. Unfortunately, however, the animal studies done both in the Soviet Union and the United States have relied almost exclusively on the use of rats or other rodents for experimental subjects. While such studies may be of value to those interested in the welfare and behavior of rodents per se, they can only be of marginally suggestive relevance to the effects of pollution on man. This is because of the fact that the mechanisms employed by rodents to respond to environmental stresses are very different from those employed by man. For example, the respiratory systems are quite different, as are the general temperature regulatory mechanisms. Indeed, it has been demonstrated that the biochemical systems involved in the hypothalamic regulation of temperature are qualitatively different in rats, cats, and monkeys (Feldberg, 1968). A growing literature in pharmacology and toxicology is also demonstrating the fact that there are qualitative differences in the way in which various species respond to various exogenous substances, so that exposure to a substance which is lethal to rats may be without effect in man, and vice versa (Wolstenholme and Porter, 1967).

One partial solution to these problems is to begin intensive studies on the behavioral effects of air pollutants on a species much more similar, physiologically, to man--i.e., monkeys. In addition, in order to approach the problem of realistic standards described above, the behavioral observations should involve tasks considerably more complex than simple sensory threshold determinations or performance on a simple reinforcement schedule. Such tasks might include choice reaction times, delayed match to sample (a procedure involving short-term memory), or complex sensory discrimination problems which yield relatively stable but sensitive behavioral measures, and which more closely approximate situations encountered by man, such as driving or typing, that require maximal behavioral efficiency. Studies of this type can be run on monkeys in polluted atmospheres for periods of two or three years without major difficulty (Reynolds, 1970). It is likely that many of the major pollutants affect behavior as much by altering metabolic processes as by acting on sensory receptors. In view of the fact that changes in ambient temperature can also influence metabolic processes, it would be important to establish the effects of various air pollutants on behavior under different temperature conditions.

Because there are toxicological differences even between man and monkeys--as well as between different men--the degree to which such monkey data can be generalized to man should be ultimately confirmed. This can be done, to some extent, by comparative biochemical, toxicological, and pharmacological studies. In addition, however, the data derived from the monkey behavioral studies should be used to specify parameters for precise short-term behavioral tests on man, which could then be conducted with a minimum of hazard and discomfort. Such tests should not only be measures of individual performance, but also of behavioral efficiency in a social setting, as there may be important interactive effects of pollution and social stimuli.

Data derived from such a program should provide a realistic basis for standards for maximum acceptable concentrations of air pollutants, which although below the tissue pathology criteria, are still derived from demonstrable deterioration of performance. The persuasive force of such standards can easily be related to socially, politically, and economically important issues such as safety and work efficiency, aside from any humanitarian considerations. It is unfortunate, but such a demonstration of an economic quid for the quo of relatively clean air may be one of the few practical inducements in bringing about the reduction of pollution standards in this country to a level much below the point where children are already not permitted on playgrounds on an increasing number of days each year, and an increasing number of people are forced to leave certain of our cities in order to stay alive because of the toxic effects of air pollutants.

TASKS RECOMMENDED FOR AIR POLLUTION RESEARCH

We would recommend three basic tasks: First, the effects of long-term exposure of monkeys to the principal air pollutants (ozone, nitrogen oxides, and carbon monoxide), on choice reaction time, delayed match to sample, and visual discrimination. Second, there would be the addition of different levels of ambient temperature as a variable in the procedure above. And, finally, short-term studies might use data derived from the two tasks above on the effects of pollutants and temperature on human performance.

CHAPTER

4

**POLITICAL SCIENCE
AND AIR POLLUTION:
A REVIEW
AND ASSESSMENT
OF THE LITERATURE**

Ronald O. Loveridge

INTRODUCTION

The control of air pollution requires government intervention; thus, the problem of clean air becomes a political issue. Political authorities must decide how much, what kind, and where air pollution should be controlled. Many observers agree that clean air depends more on political decisions as to money and power than on developments in science and technology. Yet, despite the severity of air pollution and its political context, little specific research has been completed to date by political scientists. However, within the last several years, political scientists have shown an increasing interest and competence in policy analysis. When applied to air pollution, policy analysis can identify a number of general obstacles to pollution control, including the structural and cultural features of the polity, the relative strengths and weaknesses of the policy participants, and the varied and complex problems presented by the policy process. If the policy process is divided into the stages of formulation, legitimation, and implementation, certain research priorities can be identified. First, we have little information about the responses and demands of the public in terms of air pollution. The case is made that we should have detailed and objective information on the attitudes, views, and responses of the general public. Second, perhaps the key to air-pollution control is the compliance process, where emission standards are enforced. We have very little information on the effectiveness of control agencies, especially in the control of stationary sources. The case is made that we need field research on the administrative agencies which apply the technology of control. For example, how do differences in organization and political support influence the setting and enforcing of air pollution-standards? And third, beyond immediate problems of compliance, future projections by some scientists of pollution catastrophes suggest that possible new institutions, controls, and incentives should be examined and evaluated. Even if such forecasts of doom are only partially true, we need to investigate new ways of maintaining tolerable air for the residents of the nation.

ISSUES AND ANALYSIS

Political Science and Air Pollution Research

The pursuit of tolerable air for U.S. residents has led many observers to conclude that control depends on the "great game of politics." Solutions, they say, depend more on political choices and decisions than on developments in science and technology. Control efforts--if more than symbolic--require commitments of resources and regulation of activity that can only be provided by direct and concerted intervention of government. Without government intervention, control efforts have been largely ineffective and minimal. Yet, government intervention in itself does not insure clean air. For example, California has attempted to control air pollution since the early 1950's; the state and its major counties have passed numerous intervention measures. Nevertheless, the pursuit of air pollution control returns again and again to questions of government intervention and the political process (the formulation, legitimation, and implementation of legislation).

If air pollution control is largely a political question, what does a review of the literature tell us? The first problem is--What literature? Air pollution has been a popular topic in general editorial magazines and major newspapers. Moreover, the National Air Pollution Control Office (NAPCO) will send boxes of materials on request. Further, does every speech by a politician or control official qualify as part of the political literature? Despite the fact that probably the best ideas and hypotheses can be found in these disparate sources, we decided to search only for those works completed by accredited political scientists. With this criterion, the literature available is radically reduced.

Political scientists have focused their research efforts on the policy process and not policy problems. Although problems such as poverty and war have been studied, strikingly few political scientists have devoted their attention to environmental problems, much less to air pollution. Walt Anderson (1970), edited a reader that is noteworthy because not one of the twenty-seven selections--except for a selection by the editor--was written by a political scientist. Nor do we know of any political scientist who can be called a resident expert on air pollution. In fact, aside from a speculative essay or two, we have found only four research attempts of any consequence on air pollution. And, when closely evaluated, the only work that stands out as useful is that by J. Clarence Davies III (1970). His text serves as an excellent although general introduction to the problems of air and water pollution and the policy process. Therefore, what we, as political scientists, know about the politics of air pollution in any specific sense is small indeed. The problems of air pollution have thus far not provided the occasion for testing or applying political concepts and theories.

Nevertheless, the conspicuous lack of research on air pollution does not mean that political scientists are ill equipped to make a significant contribution to pollution-control efforts. Probably the most widely accepted definition of what political scientists study is offered by David Easton (1959: 129-148): "The authoritative allocation of values for a society." We have developed an extensive literature on the policy-allocation process and who gets what, when, how, in terms of policy values. It is these assumptions, approaches, hypotheses, and and findings that offer research promise to the question: How can the nation formulate, adopt, and implement effective yet equitable and realistic air pollution control measures? Given the objectives of clean air, political scientists can study questions dealing with the characteristics and demands of the public, interest groups, the legislative process, and the administrative process. Moreover, political scientists are beginning to undertake serious evaluation research and, more generally, to investigate policy content as well as the policy process.

Policy Analysis

The discipline of political science has been distinguished by two approaches to the study of politics, traditionalism and behavioralism, and perhaps now witnesses the emergence of a third, postbehavioralism. The traditional study of politics was descriptive in method and focused on normative and institutional questions about the political order. The traditional approach fostered major developments in the area of public administration. Following World War II, a new emphasis, behavioralism, sought to bring rigor and science to the study of politics. The empirical focus was individual behavior. Extensive research efforts were devoted to socialization, voting, and like inputs to decisionmaking. In the late 1960's, the behavioral approach came under attack by critics concerned with the reluctance, even unwillingness, of the discipline to examine pressing societal problems. These critics advocated a shift in resources from basic to applied research,

for, they said, the problems of war, race, urban decay, poverty, and
pollution mandated immediate research attention. The results have
been the tentative beginning of a third approach, postbehavioralism,
which stresses the need to apply the tools and theories of political
science to the critical problems of our times. The new approach, in
substance, calls for policy analysis and engagement by political
scientists.[1]

No field in political science has, in fact, received as much recent
attention and interest as policy analysis. Since 1967, a concern with
policy has moved from the periphery to near the center of our research
activities. For example, in urban politics, we now ask not who governs,
but, instead, what differences in policy it makes who governs. The
purpose of policy studies is description, analysis, and understanding.
And, as Ira Sharkansky (1970: 2) pointed out:

> The more we understand the policymaking process (that is,
> the more accurately we can describe it, account for differ-
> ences in policies chosen, or differences in the effects of
> policies), the better we can inform those who would make
> suggestions for change.

Intensive work is being undertaken in terms of the development of
basic policy concepts, measurements of policy, determinants of
policy, and evaluation of public policy. The study of public policy is
now seen as one of the obligations as well as a proper subject of the
discipline.[2]

Contributions to Solving the Air Pollution Problem

Let us assume that air pollution can be brought under control
with existing technology if money and political power are provided.
What, then, can political scientists contribute in the way of applied
research? The specific utility of political science rests with policy
evaluation and policy recommendations. First, we can examine the
present policy process to see what has happened to the money (or
lack of it), and political power (or lack of it), in past legislative
measures at the county, state, and federal levels. What difference
has past legislation made and why? Why have some measures
achieved at least limited success while others have failed? What
role do citizens, interest groups, legislators, and administrative
officials play in the present policy process of air pollution control?
And, second, based on examination of what now happens and drawing on
findings and tools in our literature, we can propose legislative options
as well as recommend new arrangements to reach agreed policy
objectives. Or, more broadly, we can study the economic, social,
and political steps necessary to obtain policy goals. In many ways,

POLITICAL SCIENCE AND AIR POLLUTION 49

we lack not reforms but rather sound and informed policy strategies
to initiate, adopt, and especially to implement control-reform recommendations.[3]

Politics of Air Pollution

For several reasons, air-pollution control is a political problem.
Control is unavoidably the responsibility of government, for the incentives of the private market encourage rather than abate air pollution.
Major steps by the private sector to control air pollution that are unrelated to actual or anticipated government intervention are infrequent
and isolated. The requirement of government intervention leads to
a number of crucial political questions. Political authorities must
decide how much, what kind, and where air pollution should be controlled. Resources are scarce, and air pollution must compete with
other problems for money and political power. More specifically,
political authorities must decide what will be the air-quality goals,
ambient-air standards, emission standards, and, perhaps most importantly, how closely the emission standards are enforced. Problems
of compliance become especially important, as Davies (1970:202)
noted:

> Compliance is the most vulnerable part of the pollution-
> control process, the stage at which failure is most likely.
> It does not cost anybody to write good laws or set strin-
> gent standards if the laws and standards are not enforced.

Efforts directed at improving the quality of air would seem to require,
therefore, at least some understanding of the politics of air pollution.

Policy Response from the State of California

Before presenting an overview of the politics of air pollution,
it would be remiss not to stress the monumental efforts of the State
of California and many of its urban counties to control air pollution.
California has been the national pioneer, at the forefront in research,
standard setting, and compliance. The Los Angeles Air Pollution
Control District has a worldwide reputation for its success in controlling stationary sources and, as a result, is a much copied model
in terms of organization, standard setting, and enforcement procedures.
Numerous other counties have control districts, some fairly effective,
others not. The State Air Resources Agency is the most important
state-control agency in the nation, partially for its over-all authority
over stationary sources, but, more importantly, for its regulation of
moving vehicles. The California Air Resources Board can set more
stringent automobile-emission standards--thanks to the Murphy

Amendment to the Clean Air Act of 1967--than those of the U.S. Department of Health, Education, and Welfare (HEW). Because no other state has this power, California sets, in effect, the standards for the nation. The Technical Advisory Committee to the Air Resources Board is a major repository of scientific expertise on air pollution, rivaled only by special committees that NAPC can summon. More research on air pollution has been conducted in California than in any other state. Many politicians and control officials have worked diligently and sometimes with more than modest success in passing the legislation necessary for implementing air pollution control. The public has long been aroused and has supported what air pollution legislation has been adopted. The question then arises: If California leads the nation in air pollution research, standards, and compliance, why does California continue to have a serious air-pollution problem? Or, put into the context of this chapter, what political reasons account for the apparent lack of results?

Participants and Problems of the Policy Process

To call attention to the major characteristics of the politics of air pollution, we will draw at length, for the remainder of this section, from a paper that we delivered to the First National Symposium on Habitability.[4] This commentary should answer the question as to why control efforts--although saving such areas as Los Angeles and perhaps San Francisco Bay from becoming uninhabitable cemeteries-- are frequently thwarted by the policy process. While the focus is on the issue of the general environment, most of the major points also apply to the specific issue of air pollution politics.

Policy Process in the United States and the Issue of the Environment.
We will highlight several structural and cultural features of U.S. politics and then, in more detail, we will examine the major participants on the issue of environment. This section makes no pretense of being a primer on U.S. politics. Instead, we will selectively comment on political facts of life that can or will importantly effect environmental policymaking.[5]

Structure and the Policy Process. U.S. public policy is formulated, adopted, and applied within the context of social pluralism and constitutional fragmentation. The deliberative choice of environmental policies takes place in a political system characterized by decentralization, pluralism, bargaining and coalition building, and presence of veto groups.

> The making of governmental decisions is not a majestic march of great majorities united upon certain matters of basic policy. It is the steady appeasement of relatively

> small groups. [The U.S. polity] is a markedly decentralized system. Decisions are made by endless bargaining; perhaps in no other national political system in the world is bargaining so basic a component of the political process. [The U.S. polity] does provide a high probability that any active and legitimate group will make itself heard effectively at some stage in the process of decision [Dahl, 1956: 150].

The result is that policy change is almost always incremental. What is feasible is only incrementally or marginally different from existing policy, for drastic policy changes fall in most instances beyond the political pale.

An important corrolary to a politics of social pluralism and constitutional fragmentation is that groups do not possess the same amounts of resources to influence policy choices and decisions. In particular, no one should confuse public opinion with effective opinion:

> To some degree the difference between immediate (effective) and remote (public) opinion is that between organized and unorganized opinion. The alert and sensitive opinion is likely to be organized into formal groups whose spokesmen are on hand when Congress and state legislatures are considering action. They make their views known in the executive and administrative offices and to the public through the media. Remote opinion is less likely to be so organized. It has no lobbyists, although politicians may give prayerful consideration to the nature of that opinion and to the probability that it will find expression at the next election [Key, 1961: 429].

On environmental matters, it is likely that immediate interests will vanquish the public, for in no way can citizens match the organization or resources of U.S. economic associations and corporations. (In an excellent study of the adoption of the major domestic programs of 1960's, James Sundquist (1968: 506-37), points out that very few measures passed without the organized efforts of powerful national interest groups.) Up to now, group bargaining has in fact ravaged most comprehensive control standards and environmental programs.[6]

<u>Culture and the Policy Process.</u> Policymaking is not conducted in a normative vacuum. Rather, policies are formulated, adopted, and applied within powerful cultural constraints, that is, conditions and values not susceptible to much shortrun change. These normative constraints guide the choices of policymakers by requiring them to proceed in certain ways and by opening or closing certain policy

options. As to the environment, several values seem important in setting limits and directions for policy choices. In asking why we are such giants in production and such dwarfs in creating public amenity, Edmund Faltermayer (1968), cites four fundamental attitudes--belief in individualism, disrespect for nature, assumption of the inexhaustability of the nation's resources, and a suspicion and skepticism of planning. If we accept these four as significant and add the concepts of growth and progress, the issue of the environment confronts value commitments that will act to tether large-scale policy innovation and change. Before we can propose environmental policy programs, we must take stock of our U.S. culture and ask questions about the good life and the good society and, perhaps more important, what cultural values are now framed in the principles and practices of our polity.

<u>Participants and the Policy Process</u>. Within structural and cultural constraints, the policy process can be approached by examining its principal participants, for the participants will largely determine the score and scope of environmental policies. Brief portraits of seven participants--general public, attentive public, interest groups, media, scientists, government agencies, and elected officials--will be offered. These portraits are, of course, incomplete and highly selective; nevertheless, they attempt to identify characteristics central to the environmental crisis and public policy.

First, let us consider the general public. Writers on pollution of the environment have often advanced a contention similar to that advanced by Howard Lewis (1955: 220): "Fair, sensible, practical means of controlling air pollution are at hand. Their implementation requires, first, before anything else, the activity of an aroused citizenry." The great hope of U.S. reformers, conservationists or otherwise, has been the general public. As a result, writer after writer has called for information campaigns to alert, educate, and involve the public. Although the causal reasons are unclear, the call for an aroused public was answered in 1970. By then, the issue of the environment was the primary policy focus of many citizens. Moreover, polls indicated that the public was willing to pay for improving the environment. In fact, it is difficult to envision a more aroused and unanimous public than now exists on the general position of improving the quality of the environment (Loveridge, in Atkisson and Daines, 1970: 58-63).

While the response of the general public has set in motion the activities of others, we should not translate public opinion as effective opinion. The general public is not a deeply persuasive, organic force. Instead, it consists of citizens who are neither well informed, nor deeply motivated, nor particularly active, even on the issue of the environment. The citizen <u>as</u> citizen is largely an impotent partisan, for he lacks the expertise and resources to gain access for influence. Dorwin Cartwright (1964: 227) aptly stated:

> Surely, the citizen who doesn't belong to a strong organization with a lobby in Washington or who doesn't know a Congressman's wife has little chance of influencing governmental policies and procedures except through rare visits to the voting booth.

Even if highly motivated, most citizens lack the wit and means to make the case for specific environmental policy programs. In essence, the general public cannot be expected to exercise innovation or take leadership to achieve a habitable environment.

Second, let us consider the attentive public. Students of public opinion frequently make the distinction between an attentive public and the general public. The attentive public is those citizens with some detailed information and interest in policy matters, perhaps 10 percent of the adult population (Rosenau, 1961). On the quality of the environment, they share views similar to the general public, except that their response is more intense and more differentiated. The attentive public shows every indication of becoming increasingly aware of and concerned with the question put by President Richard M. Nixon in his 1970 State of the Union Message: "The great question of the 70's is: Shall we surrender to our surroundings or shall we make our peace with nature and begin to make reparations for the damage we have done to our air, to our land, to our water?" Thus, the attentive public promises to be a critical and demanding audience to the speeches, controversies, programs, and results of environmental policies and priorities.

However, the attentive public also shares the same weaknesses as the general public in terms of their relationship to the policy process. They lack the access, expertise, information, and even resources directly to influence environmental policymaking. Not organized into formal groups, they essentially react to media cues and messages. Moreover, they remain confused as to what action they can take or what environmental policies are practical and significant. In addition, the arenas where major environmental policies are formulated and adopted are remote from and even alien to their community and work experiences--as policy moves from the city level to county, to state, to the federal level, it becomes increasingly difficult for the attentive public to exert direct influence. Nevertheless, the role of the attentive public as media consumers, and, in turn, opinion leaders, should not be discounted, for they importantly structure the direction and tenor of the images by the public and the politician regarding environmental problems and priorities. Probably the most crucial question is: How long will their focus on the environment be sustained? Several years ago, Daniel Moynihan remarked that the white middle class had become saturated with and had lost interest in black-white relations. Will the same process occur for the attentive public on the issue of the environment?[7]

Third, let us consider interest groups. The most widely used and accepted perspective of U.S. politics has been the group approach. V. O. Key (1964: 17) offers a representative explanation:

> At bottom, group interests are the animating forces in the political process; an understanding of American politics requires a knowledge of the chief interests and of their stake in public policy. The exercise of the power of goverance consists in large degree in the advancement of legitimate group objectives, in the reconciliation and mediation of conflicting group ambitions, and in the restraints of group tendencies judged to be socially destructive.

In the arena of environmental politics, four kinds of interest groups merit evaluation; they are single issue, conservation, professional, and economic groups. To no small extent, the activities of these groups will determine who gets what, when, and how, in environmental policies and programs.

In increasing numbers, ad hoc citizen groups have organized around various environmental issues. For example, in Southern California, many groups have risen to smite the dragon of air pollution: Stamp Out Smog (SOS), Clean Air Now (CAN), Group Against Smog Pollution (GASP), Clean Air Council, and so on. These groups are sometimes important in initiating controls and supporting local programs; however, they usually do not become, in any sense, popular movements. Perhaps the most troublesome problem for single-issue organizations is to provide continuing and satisfying incentives for their membership. As a result, these groups are noticeably ineffective in sustaining a concerted, partisan influence on environmental policies and programs. While there are notable exceptions, single-issue groups tend to be used for support purposes by control agencies and to be ignored or disregarded by most politicians.

While nearly every county in the country has some sort of citizen conservation group, these groups are not formed to protect the whole environment, but, rather, to battle a specific threat quite close to home. Although impressive in numbers and intensity, they do not possess the resources necessary to be powerful legislative advocates at the state or national level.

> Individuals, acting on their own, cannot generate enough
> pressure to get the legislation that is urgently needed now;
> strong, continuous, sharply focused pressure is needed.
> Amazingly, there are almost no citizens' organizations any-
> where in the United States that are capable of applying this
> kind of pressure. . . . [Faltermayer (1968: 213-214)]

Especially in Washington, D.C., the national conservation groups are largely ineffective and unknown. While conservation groups sometimes win on specific issues, they cannot be expected to take the full range of environmental problems to the people and their legislators and political executives. The lack of staff, loose organization, and divergent interests and priorities will prevent them from assuming the national leadership necessary to achieve a more habitable environment. (It is a useful exercise to ask what such as groups like the Izaak Walton League, Ecology Action, Wilderness Society, Zero Population Growth, and the National Wildlife Federation have in common.)

Professional, good government, service, and even social groups have also taken up environmental causes. The resources of these groups in the form of numbers, expertise, reputation, and organization represent major potential pressures for environmental reform. For example, during the summer of 1969, in Riverside, California (a city of 140,000 population, about 50 miles east of Los Angeles), the State of California's acceptable ozone level was exceeded for sixty consecutive days. Many city groups reacted by attempts to use their local and statewide organizations to work for air pollution policy changes and reforms--the groups included the Tuberculosis Association, Medical Association, League of Women Voters, Junior Chamber of Commerce, Junior League, City Council (California League of Cities), and Board of Supervisors (California Supervisors' Association). When such groups are combined with the conservationists, the resources available to influence environmental policy decisions are dramatically enhanced. The crucial problem with such a coalition lies in the kinds of policies, if any, these diverse groups can agree upon and work for. Nevertheless, it seems appropriate to note the increasing prominence of professional and other groups who have now entered the arena of environmental politics.

The most common basis for group action is economic. And, the corporate groups, with economic roots in pollution and the present market structure, have a near choke hold on the group resources to influence environmental policy. If politics could be compared to a billiard ball responding to the strongest group pressures, the outlook for significant environmental change would be dim indeed. The general response of most corporate groups is to delay if not oppose environ-mental policies that will present new or additional costs to their business. At present, incentives are high to lobby against change. For example, Benjamin Linsky, former Chief Administrative Officer for the Bay Area Air Pollution Control District, explained: "For every year an industry can defer spending $100,000 to install air pollution control equipment, they can usually keep $25,000 more in their pocket before taxes" [quoted by Cox, 1969]. Corporate groups have consistently demonstrated that they will not take remedial action to control pollution or to change marketing and resource habits unless compelled by government. However, business is now aware of and reacting to the growing public concern over the environment.

> Advertising and public-relations departments across the country are hard at work telling us how much corporations care about the environment they are befouling. In full-color, two-page ads in fancy magazines, oil companies inform us how the fish actually like detonations for underwater wells, or that their brand of gasoline will pollute the air just a little bit less [Margolis, 1970: 124].

Nevertheless, what is good for U.S. business is not yet good for environmental policies or priorities. The major corporate interests loom as primary obstacles to environmental victories in the legislative or regulatory arenas.

Fourth, let us consider the media. Some observers see the United States moving from a group-based politics to a media-based politics. In terms of setting issues on the national agenda and providing them with popular momentum, the role of the media is difficult to overemphasize. The media tell us what to think <u>about</u>, and, in many ways, what to think, period.

> The media . . . structure a very real political environment which people can know about only through the media. Information about this environment is hard to escape. It filters through and affects even persons who are not directly exposed to the news or who deny that they are paying a great deal of attention to what the media say. There is something pervasive about the content of the mass media, something that can make its influence cumulative [Lang and Lang, 1968; 305].

It is difficult to escape the attention now given by the media to environmental matters. Countless examples could be cited, but let us conclude by saying that the media are devoting unprecedented time and space to the quality of our environment. The principal weakness of the media is its tendency to move from one dramatic topic to another and thus fail to detail and sustain its coverage. We must ponder how long the environment will remain an issue if the media no longer regards it as newsworthy.

Fifth, let us consider scientists. Michael Reagan (1969: 23), pointed out that "science has developed the capabilities by which man can in startlingly large measure re-make (or destroy) his environment, his social structure, and even himself." Because scientists possess these capabilities, they can play a major role in the deliberative process on environmental policy choices. In speaking out on the environment, however, scientists are manacled by three drawbacks. For one, as Rene Dubos explained (1970: 22): "We cannot solve the environmental problems of our society because in most cases knowledge of the interplay between man and his environment is either inadequate or totally irrelevant." Perhaps more important, little coordinated and integrated research is being mounted on the environment. Most research takes the form of scattered and fragmented projects by individual scientists. And, third, a breakdown in communication exists between most scientists and politicians. In interview after interview, politicians have commented on the failure of scientists to present findings in language they can understand, the tendency to offer conflicting and highly individualized interpretations, the continuous plea for more research money, and the seeming concern with professional rewards rather than policy reforms. Unless scientists can alter these drawbacks, their expertise will remain outside the partisan mainstream of the environmental policy process.

Sixth, let us turn to government agencies. Francis Rourke (1969: vii), noted that "it is in the crucible of administrative politics today that public policy is mainly hammered out, through bargaining, negotiation, and conflict among appointed rather than elected officials." Much of policy innovation and leadership is centered on government agencies, for, in addition to formal authority, they have a near monopoly on technical information and expertise. Not surprisingly, these government agencies share a major responsibility in promoting policies, and in turn, in organizing a political basis for their survival. As a result, agency "withinputs" are crucial to improved environmental policies and programs. Yet, these same government agencies present some profound weaknesses as champions of environmental reform.

There are five weaknesses that call for specific comment. First, there are many government agencies that lay claim to some portion of the environmental crisis. Each of these agencies has a limited mission, with no one--except perhaps the President's Council on Environmental Quality--responsible for the whole environment. Attending to the environment requires more than piecemeal action by federal agencies, many of whom compete with (rather than cooperate with) one another. Second, although government agencies can exercise considerable discretion in advising on and implementing policies, they are constrained by the rules of the game from making a policy case very much different from that of their political bosses. Especially on controversial matters, government agencies face important limitations on the kinds of bargaining and symbolic leadership they can exercise. Yet, before much in the way of improvement occurs, controversial policies have to be promoted. Third, spirited competition exists among various levels of government on pollution and environmental matters, with cities, counties, states, and the federal government jealously guarding their own programs and prerogatives. For example, the principal weakness of the Air Quality Act of 1967 is the provision that states and cities have the responsibility to set up effective standards and regulatory agencies. While the actual results--as anticipated by national control officials--have been almost wholly disappointing, the provision was necessary to solicit approval from the state representatives in Congress.

A fourth weakness is, more often than not that administrative agencies tend to provide reassuring symbols for the public and concrete benefits for organized interests. Most administrative agencies have few public friends; instead, they deal almost exclusively with clientele groups who they are set up to regulate--namely, those industries that manage the economic resources of the nation. Yet, administrative agencies can survive only so long as they continue to secure the support of politically effective groups, and, through these groups, to secure legislative and executive support. As a result, Edelman (1964: 23), noted: "Tangible resources and benefits are frequently not distributed to unorganized political interests as promised in regulatory statutes and the propaganda attending to their enactment." Moreover, a strong tendency exists for group-agency relationships to be highly resistant to disturbing changes. Fifth and last, it is difficult to identify any government agency as being especially effective as a promoter and implementer of environmental reform policies. The results, at best, have been creeping incrementalism, and, at worst, business as usual. With few exceptions, government agencies are unwilling or unable to make important sallies at established ways of doing things.[8]

Finally, let us consider elected officials. Unlike other participants, elected officials have the formal responsibility to set priorities, bargain for desired policies, and, most importantly, legitimate policy decisions. In addition to making binding decisions, elected officials now find it necessary to take positions on and often to promote environmental policies and programs. What this means, however, is still uncertain, for everyone declares against pollution and for improving the quality of the environment. The crucial question is whether the political rhetoric will be matched by political action. Will the political executive devote the time and energy to follow up his good intentions? (After President Nixon's special message on the environment was

POLITICAL SCIENCE AND AIR POLLUTION   59

sent to Congress, Newsweek (1970) quoted a top Senate environmentalist as saying: "This is only round one. The real game will be played in the dark corners of the federal agencies and in legislative conference." The Presidential pressures that are brought to bear for strict regulations and satisfactory funding will be the most important measure of President Nixon's interest in environmental matters.) And will the legislators expend the resources necessary to fund programs and enforce rules? Again, we must ask how long political interest among elected officials will continue when programs do not achieve immediate results and the national audience becomes exhausted by media messages and stump rhetoric. Further, pollution control is perhaps the easiest step. The more complex issues, such as the use of land and changing-market incentives, will require a commitment to radical innovation and leadership that few elected officials yet seem willing to consider.

Policy Problems and the Issue of the Environment

Philip Berry (1970), President of the Sierra Club, set down three steps in conservation achievement: First, the public must be persuaded, then the politicians must take up the cause, and, finally, conservation rhetoric must be matched by political action. Steps one and two have largely been met. These two developments are of fundamental importance, for the issue of the environment is now on the public and political agenda for the nation. Unfortunately, alarm does not put out fires. No matter how widespread the agreement on improving the environment, these hopes and promises have to be translated into policies. And here the problems begin in earnest.

To examine environmental matters and the obstacles to reform presented by the policy process, we have adopted a classification scheme offered by Mr. Kent Jennings (1964: 107-9). The policy process, says Jennings, can be divided into five steps: initiation of action, fixing priorities, utilizing resources for gaining acceptance of chosen alternatives, legitimation, and implementation. This method of classification provides five relatively distinct points from which to analyze the relation of environmental matters to the policy process.

Policy Initiation. Policy initiation is more difficult to execute than it appears. A problem must be recognized and defined; then, possible policies must be formulated and evaluated. While seemingly simple to execute, these tasks become highly troublesome in practice. Conflict and controversy usually develop first over what the problem is and then what kinds of policies can and will work. And, clearly, the more complex the problem, the more difficult becomes its definition and resolution.

If nothing else, the environmental crisis defies much more than a limited understanding of its causes and consequences. For example, let us look at the problem of air pollution. While laymen and policymakers evince confusion, scientists likewise show no lack of uncertainty or disagreement. Although it is frequently said that financial cost and political will are the major obstacles to air-pollution control, Rene Dubos (1970: 20), wrote: "But, in fact, even if we had limitless resources we could not formulate really effective control programs because we know so little about the origin, nature, and effects of most air pollutants." Or, when trying to assess the urgency of control efforts, we are dismayed at the conflicting scientific claims as to whether or not the world is coming to an end. And if we remain confused, we are certain that the public and its policymakers must share a similar sense of bewilderment. If the questions on the causes and effects of air pollution are then expanded to the total environment, the definition of the problem poses obvious, enormous difficulties.

The complexity of the environmental crisis would not pose such a difficult policy assignment if the environment represented a single problem. Instead, the environmental crisis is a shorthand expression for a host of separate as well as interrelated problems. For example, Charles Abrams (1968: ix), offered this itemized description:

> We have been wanton in our destruction of nature's gifts. Much of what we have built is either slum or potentially slum. Noise, litter, vandalism, violence, pollution, monotony, crowding, and sprawl describe our urban places. The world's richest nation has the poorest of environments.

While perhaps these problems are not individually awesome, the cumulative effect of such a catalogue of environmental ills should underscore the difficulty in deciding on effective policy approaches. Beyond isolating problems, it has been frequently pointed out that the earth should be regarded as a spaceship and that life is dependent on certain fragile functions being sustained. If we accept this conception, it becomes even more important to approach the environment not in terms of discrete problems but rather with some over-all design. But the crucial questions remain: Where do we begin? How do we begin? Where do we want to go? Will so-called solutions create even more intractable problems? And so forth. The complex character and interrelated nature of environmental problems present policy innovators with a crowded yet confused agenda of policy choices and priorities and with costs and benefits mostly obscure.

<u>Fixing Policy Priorities.</u> After policy responses to the environment are initiated, policymakers must decide what priorities should be assigned to different problems and possible solutions. It would be nice if we had an unlimited public treasury, expertise, and manpower,

as well as the ability to satisfy sundry private desires and public wants. However, alas, the resources of our society are scarce and have alternative uses. In fact, scarcity and alternative uses are probably the crucial postulates of political life and thus at the source of most policy conflict and disagreement. Policymakers--on the basis of demands, knowledge, tradition, and/or values--have to reach some agreement on what is important, what can be done, and what should be done immediately and in the future. The priorities they establish, even if by default, determine what resources at best will be allocated to improving the quality of our environment.

There are no easy answers and thus no ready agreement on what policy priorities should be set for the environment. Several troublesome questions confound a consensus. First, what environmental problems require immediate attention? And, especially, what problems require investment of large amounts of resources in order to avert serious threats of life itself? Here the disagreements begin, for partisans offer various claims. Second, assuming agreement on what problems are of immediate importance, what resources then should be devoted to less "threatening" but no less serious environmental problems? Can we afford to preserve life without attention to its quality? And, third, many other alternative uses can also make strong claims on available resources. How can environmental needs be balanced against social and economic needs? For instance, some observers say poverty and racial discord are more important than the environment. Others still see the major problem as the inequitable distribution of income.[9] And so forth. In short, reasonable men of good will are likely to disagree on the specific priorities of allocating resources to resolve the nation's environmental problems.

One major reason underlying the priority disagreements has been the lack of any clearly specified objectives. It is one thing to talk about achieving a habitable environment and quite another to identify concrete and measurable goals. And as the policy focus moves from pollution to environment to quality of life, policy criteria become increasingly important.[10] Without established guidelines, our environmental policies will be largely shaped by immediate, situational variables that take the form of putting out one brush fire after another. Yet, nowhere are we familiar with any major study or report that has presented environmental objectives in a manner upon which priorities can be formulated. Because policymakers depend on certain, ill-defined, and often contested criteria to frame and propose environmental policies, we would expect disagreement on specific environmental priorities to be the rule rather than the exception. However, even if a Presidential commission could establish some widely accepted objectives, the task of fixing environmental priorities would be only somewhat less difficult. For, no matter what the costs and benefits of various objectives, the new priorities must compete with present values and ways of doing things.

Policy Acceptance. Once policy priorities are set, the political game of bargaining and conflict begins. Policymakers must combine diagnostic and strategic considerations. In particular, they must marshall sufficient leadership, money, support, appropriate symbols, relevant information, and, in addition, pursue intelligent strategies. Without these efforts, attempts to secure agreement on significant proposals for policy reform will be inordinately difficult. No matter how much a change may be valued, specific policies must be won over opposition. And, on environmental matters, the scope of major changes will directly impinge on the vital interests of many groups and individuals, most of whom will have high stakes in the status quo.

The stakes for delaying if not opposing major environmental reform are high, for it is likely that proposed changes will directly conflict with the interests of many political institutions, economic groups, and individual preferences. Separately or together, this opposition can indeed be formidable. Political institutions are jealous of their own prerogatives, wedded to established ways of doing things, and ill equipped to carry out comprehensive planning. The local, regional, state, and federal bureaucracies are more friend than foe, with sustained coordination rarely achieved on major policy matters. Efforts to reshape the political structure to direct national growth and coordinate environmental programs will evoke opposition from the many separate, governmental bureaucracies. Economic groups are likely to oppose environmental proposals for quite obvious and logical reasons. Many proposals will clash with the market incentives, profit objectives, or other economic ideas of U.S. corporate interests. While perhaps endorsing governmental action in the abstract, strong opposition can be expected on specific environmental policies. And, finally, the public has not given a carte blanche mandate to policymakers to pass environmental proposals. Especially to do more than underwrite indirect costs, the public will have to be convinced. So policymakers face the prospect of potential latent disagreement on the part of the public. And if the public is actively opposed, policymakers will find it most difficult to gain acceptance of chosen objectives. The moral is that considerable time, money, energy, and skill will have to be used to secure a coalition large and effective enough to overcome the probable opposition lineup on major environmental proposals.

Policy Legitimation. Policy measures are formally approved in the legislative arena. In most state legislatures and certainly in Congress, power is widely dispersed. As a result, bargaining, coalition building, negotiation, and compromise are again the rules of the game to adopt controversial legislative proposals. Rather than a single-majority coalition, many majorities must be put together in committees, subcommittees, rules committees, floor, both houses, conference committees, and so forth. With this decentralized structure, controversial legislation is difficult to pass, as intense minorities are provided with a number of access points. In short, the process by which a bill becomes a law presents major obstacles to anything more than piecemeal political reform.[11]

Aside from the normal legislative obstacles, passage of major environmental programs is further frustrated by the inability of the attentive public to penetrate the rhetoric of reform. The symbolic popularity of the environment often gives the appearance of action when little is taking place. Businessmen and politicians trumpet their concern for the environment--speeches are made and "achievements" advertised. With everyone on the side of nature's angels, the attentive public is presented with the nearly impossible assignment of distinguishing the good guys from the phonies. Specifically, it becomes very difficult to monitor legislative activity and to hold politicians accountable. As a result, much of the legislative cutting edge of the environmental issue is blunted.

Inasmuch as most environmental legislation is quite limited in scope, another important legislative problem faces environmental reformers. Because of the opposition of those with a stake in the status quo, even modest changes will usually require considerable outside support. Support for piecemeal reforms is, however, exceedingly difficult to muster and sustain. With this in mind, many reformers advocate Daniel Burhham's famous call: "Make no little plans. They have no magic to stir men's blood and probably will not be realized." Even big plans, however, are translated via bargaining, coalition building, negotiation, and compromise into more modest objectives. The real problem that develops for reformers lies in the ballyhoo that surrounds the passage of piecemeal changes promoted as major reforms, as Murray Edelman (1964: 26), has written: "The most intensive dissemination of symbols commonly attends the enactment of legislation which is most meaningless in its effects upon resource allocation." The unfortunate result is that the supporters will not go to the barricades time after time for the same battles. Yet, unless continued environmental reforms are adopted, the earlier piecemeal changes are likely to become Pyrrhic victories.

Policy Implementation. Some reformers assume that politics ends with the passage of legislation and that the administrative process is a one-to-one enactment of the legislative mandate. Political scholarship has almost wholly discredited this interpretation. In many ways, policy implementation shares the same characteristics as the legislative process:

> Agencies respond to group pressures by modifying existing policies or developing new ones. Bargaining or the adjustment of conflicting interests is as constant a feature of administrative politics as it is of the relations among legislators and legislative committees. Changes in policy tend to be . . . incremental in character. Bureaucrats, like legislators, are wary of sweeping innovations which may disturb existing programs [Rourke, 1969: 103].

Laws do not provide detailed rules but, instead, offer considerable room for discretion and interpretation by the administering agencies. And these laws can be repealed in effect by administrative policy, budgetary starvation, or other little publicized means.

Two structural problems, lack of political coordination and inadequate enforcement powers, merit brief comment. Environmental policy is fragmented into a host of government agencies, each with their own mission and clientele groups. These agencies often work at cross purposes and establish conflicting priorities. As single-purpose, environmental agencies are extended from the nation, to the region, to the state, to the county, to the city, to the neighborhood, the difficulties of coordination should be obvious. It is unlikely that "everyone doing his own thing" will result in significant improvement of the environment. Another issue related to the fragmentation of authority and purpose is the absence of adequate enforcement powers. Too often, reasonable legislation in terms of standards is adopted, but the enforcement procedures are such that these standards can only occasionally be implemented. Without more political coordination and viable enforcement procedures, the outlook for environmental quality is bleak.

Finally, in terms of controlling or retarding change, the influence of clientele groups on environmental agencies cannot be overstated. If the agencies antagonize the clientele groups, they risk attack and the failure of their mission. Thus, as the reciprocal relations between agencies and clientele groups are fostered, the agencies frequently mirror the interests of the clientele groups. In other words, the agencies tend to become protectors of the status quo and use their power to maintain the stakes of the groups they regulate:

> As programs are split off and allowed to establish self-governing relations with clientele groups, professional norms usually spring up, governing the proper way of doing things. These rules-of-the-game heavily weight access and power in favor of established interests [Lowi, 1969: 62].

Research and Air Pollution

Since World War II, most political scientists have focused their attention on the policy process. To review this literature would require a wholesale analysis of political-science research. Rather than undertake such an assignment, we have chosen more modest objectives. First, we will focus only on research areas that may be of direct importance to the control of air pollution. And, second, we will not inventory research findings but, instead, will discuss questions open to applied and, in some cases, more basic research. In substance, this section attempts to indicate what political science may contribute in the way of research to the control of air pollution. In order to identify and discuss research areas, the policy process will be reviewed in terms of three stages: formulation, legitimation, and implementation. This classification scheme adopts a quasi-systems model of the policy process. Formulation refers to inputs, legitimation to formal decision-making, and implementation to what happens to decision outputs. Formulation thus centers on the major supports for and demands on the political system. If supports and demands merit evaluation, what should we know about the public and interest groups in terms of the control of air pollution?

The Public

The importance of the public in the control of air pollution is incontestable. The attitudes of the general public influence and limit government intervention to control air pollution.

> After all, it is these attitudes, be they expressed as pre-ceptions, opinions, beliefs, hopes, desires, wishes, or feelings, that stably and in change, ultimately accrue into political decisionmaking and socially imposed, ordered change. These collective societal attitudes ... dictate what is feasible and necessary in society [de Groot, 1967: 247].

Decisions about air pollution will inevitably be made in the political arena where public opinion does--and should--play a crucial role. Yet, as one of the HEW criteria documents points out, "Unfortunately, very little work has been done in this particular area of measuring the response of the public to the nuisance of air pollution" [NAPCA, 1969: 7-13; emphasis added]. To discuss research relative to public responses, we will quote at some length from a paper we presented to the Symposium on the Development of Air Quality Standards.[12]

Personal Responses to Air Pollution

While air pollution specialists agree on general ways to examine the effects of air pollution, personal responses have largely been excluded from their attention or research. Few specialists would quarrel with Leslie Chambers' five types of effects: visibility reduction, material change, agricultural damage, physiological effects on man and domestic animals, and psychological effects (Chambers, 1968: 17-19). Chambers' commentary on psychological effects--the major type under which personal responses could be subsumed--illustrates how narrowly such effects are commonly defined:

> Since fear is a recognizable element in public reactions to air pollution, the psychological aspects of the phenomenon cannot be ignored. Psychosomatic illnesses are possibly related to inadequate knowledge of a publicized threat. Little effort has been directed toward evaluation of such impacts in relation to general mental health of affected groups, or determination of their role in individual neuroses. Only in practical politics has any significant action been based on recognition of the psychological attitudes induced by periodic public exposure to an air-borne threat [Chambers, 1968: 19].

This quotation, if we understand it, limits personal effects to psychopathology rather than how, for example, perceptions of air pollution change man's attitudes and actions, viz., disruption of life, work, and play habits. Yet, no matter how defined, research on "psychological effects" has been scarce indeed; for, despite an extensive literature search, few scientific studies on human behavioral responses to air pollution can be found.

Nevertheless, we do know that people living in polluted areas are aware of and concerned about air pollution. What we do not know is how perceptions of air pollution affect their attitudes and behavior. In an excellent review of public attitudes toward air pollution, de Groot (1967: 247) stressed a similar view: "We should know how perceived dangers of air pollution alter man's life style and belief systems, and psychological makeup." Therefore, we propose to explore four general types of individual adjustments to air pollution: psychological, social, economic, and political. These classifications direct attention to major personal responses that should be studied.

Psychological responses are probably the most difficult to specify, for they refer to attitudes about the impact of air pollution on the private self and perceptions of its influence on the nature and character of the human condition. Ronald Ridker refers to these responses as "psychic costs" and says they are, although overlooked, of major importance:

> This category includes everything from the anguish of death to the disappointment felt when one's view of the mountains is obscured by smog. Economists have generally ignored this category on the grounds that it cannot be accurately measured, and that in many cases its inclusion would alter the decision that benefit-cost analysis which is adequate in other respects leads to. The difficulty when considering air pollution is that there are important cases where this category would make a difference in the policy decision if it were included. Indeed, I suspect that the increased demand for clean air over the last fifty years or so comes mainly from a desire for a more beautiful environment and only secondarily from an increased knowledge of the detrimental effects of pollution [Ridker, 1966: 91-92].

Social responses refer to the effects of air pollution on an individual's life style. The individual can ignore its presence, tolerate its effects, or change his social habits. For example, various levels of pollutants could influence satisfaction with and choice of place of residence, job, recreation, vacation, and so forth. Needless to say, opinions and decisions related to these activities are central to enjoyment of life and the over-all demand for a decent environment.

Economic responses focus on remedial action available through the market place. Two kinds of responses can be distinguished. One is to avoid the effects of air pollution, e.g., purchase air-filter equipment or pollution-resistant products. The second is to treat the effects, e.g., purchase medication or special cleaning preparations. These responses are man's effort to utilize the bounty of the nation's private goods to adjust to the effects of air pollution. Beyond decisions to use one's purchasing power, other factors are also crucial to evaluating economic responses, for example, ability to pay, awareness of market choices, and so forth.

And, finally, political responses refer to an individual's knowledge, concern, and policy opinions toward air pollution and its control. Does he recognize the problem? Is he concerned? What action has he taken? What does he believe should be done? (These questions will be examined in detail below.)

Information on personal responses should have several major consequences. First, in an important sense, democratic rhetoric and policy practice can be consummated, for the public will be given at least an indirect hearing in the setting and implementing of standards. Until now, the public's perception of air pollution and the influence such perception has on attitudes and behavior has received short shrift in most policy councils. And, second, surveys can identify and emphasize personal effects that are not presently evaluated. This data should contribute to the assessment and development of air pollution control programs. As Jane Schusky, et al. (1964:1) observed:

> Public-opinion surveys are particularly valuable . . . with respect to those aspects of air pollution which are not presently capable of standardized, objective measurement. Thus, the extent to which odors, visibility effects, aesthetics, soiling, and sensory and upper-respiratory irritation become a "problem" is highly dependent upon the extent to which people see it as such, and are concerned or bothered about it. Further, the extent to which air pollution is a problem may also be reflected in the ways in which people alter their behavior in response to it. Thus, air pollution may be a factor in causing people to avoid downtown shopping areas, or certain recreational areas, or to change their residence or place of work.

Finally, surveys of personal responses should identify the extent to which the public demands and will support pollution-control measures. This reason is perhaps the most important of the three, for present evidence suggests that the public is more predisposed to control air pollution than state and local authorities. Polls have found pollution control at or near the top of priorities the public believes government should do something about. For example, a poll of California voters found, out of five issues, that voters assigned highest priority to "stopping pollution of our environment"(Pastier, 1969). If carefully documented, the urgency and intensity of the public's cry for clean air should present a charge to policymakers that cannot be dismissed or ignored.

Some Specifics of Political Responses

While exhorting the study (and importance) of personal responses, we have been unfortunately vague in identifying specific responses. Vagueness cannot be attributed to survey research, for its skills are now sophisticated, widely known, and highly scientific. Rather, there is a striking lack of relevant conceptual or empirical work on personal responses to air pollution--hard facts and completed studies simply do not exist. We do not know, in many cases, what specific responses are significant or what measures are possible and objective. Therefore, to illustrate the kinds of responses that can be studied, selected political responses are discussed below.

Political responses refer to attitudes and behavior (whether actual or latent), toward the control of air pollution. The literature on political participation says that political involvement, even in its most elementary forms, generally depends on the fulfillment of at least three conditions, according to Campbell, Converse, Miller, and Stokes (1964:97-108). First, the problem must be recognized. Second, it must arouse some strong feelings of intensity. And, third, it must be accompanied by a perception that political answers are available and that political involvement could make a difference. These conditions for political involvement provide a classification scheme to examine political responses. Thus, we will take up knowledge, feeling, and policy views toward the control of air pollution.

Political Knowledge. Political attitudes toward air pollution can be divided and explored in many ways. When we focus on data immediately relevant to the setting of air-quality standards, two dimensions seem important--first, the recognition of air pollution as a problem and, second, the perception of the personal effects of air pollution. The recognition of air pollution as a problem is the most fully studied of any question on personal responses. Whether the pollution survey is a national, metropolitan, or community effort, respondents are always asked--in some manner--if air pollution is a problem for their area of residence. The results show an increasing and widespread awareness of air pollution and a positive correlation between the levels of pollution and awareness of air pollution as a problem. If carefully constructed, a measure of awareness could be used to develop standards such as W. W. Stalker and Charles Robison (1967: 143) proposed:

> For a control program to be worthwhile, it would seem that concentrations of pollutants presenting a nuisance to as many as 50 percent of the people should be permitted for only short periods of time if at all.

While they present difficult problems of measurement, perceptions about the personal effects of air pollution also merit close evaluation. For the effects people can identify explain why air pollution is thought of as a problem. The content of these responses, by indicating what people find objectionable, irritating, or dangerous, should be instructive to those responsible for setting and implementing air-quality standards. Personal effects however, represent a wide-ranging catalogue of possible complaints. In order to illustrate, let us take one example, loss of visibility. If a high percentage of the residents of a region are distressed, perhaps angered by the reduction of visibility, the visibility statements in the criteria documents should become more important in air-quality deliberations. Moreover, a focus on visibility could have two major advantages: One, visibility is an objective measure on which we have considerable data and, two visibility could probably serve as a satisfactory indicator of other effects, many of which are premised on what people can see.

Feelings About Air Pollution. An Interlandi cartoon appearing in the Los Angeles Times pictured a sedate, middle-aged couple confronting their activist son who, with picket sign in hand, was off to another protest. With a sense of bewilderment and resignation, the parents ask, "Why war? Couldn't you protest against smog? Everyone's against air pollution." While it is surely true that most people oppose air pollution, the cartoon calls attention to the second major kind of political response: relative concern that air pollution should be controlled. The public faces a plethora of problems, all of which are competing for attention and resources. The question becomes how much they value the control of air pollution when compared to other private decisions and policy choices. Relative intensity of personal concern for pollution abatement should become a part of the policy process, for the public's priorities in addition to specific group pressures should be counted when air-quality standards are adopted and enforced.

Despite serious problems posed by comparability and intensity, personal feelings about the control of air pollution can be evaluated by survey research. Several general measures suggest ways to begin. For instance, in a 1967 national sample, Louis Harris asked whether seventeen federal programs should be expanded, kept as is, or cut back. First on the list to be expanded--ahead of only five others, including federal scholarships for needy college students and medicare for the aged--was the program to curb air pollution (Harris, 1967). In a 1968 national Gallup poll, a different measure of intensity was used to sample public opinion on the effects of environmental deterioration:

> About half (51 percent) of all persons interviewed said they are deeply concerned about the effects of air and water pollution, soil erosion, and destruction of wildlife. About one third (35 percent) said they are somewhat concerned. Only 12 percent said they are not very concerned [quoted by Cahn, 1969].

And, finally, an ingenious example of a more complex instrument to measure personal concern was developed by two students in the 1968 California Institute of Technology summer project, James Beck and Marrianna Stapel (1968); they attempted to measure "relationships between values, those relating to transportation and air pollution, and to provide meaningful uses for the information obtained." Thus, while much work has to be done, objective measures of the public's relative concern for the control of air pollution can be developed.

POLITICAL SCIENCE AND AIR POLLUTION

Policy Views. Two kinds of policy views can be explored: orientation and policy stand. Orientation refers to how the individual directs himself in terms of action. The focus would be on how the respondents have approached the control of air pollution. Policy stand is the preference for collective action: What should the "authorities" do about air pollution? A former California legislator, Byron Rumford (1966: 360) reminded us: "Those of us who make the final legislative decisions often find ourselves caught in the dilemma of seeing the problem and knowing of the solutions but lacking sufficient popular support to carry them out." Measures of policy views should provide exact indication of the kind and extent of the support that exists for air pollution control procedures and objectives.

The concept of orientation directs our attention to the public's political reaction to air pollution. In what manner or form do people participate? Stephen Ayres (1967: 15), concluded a speech with the plea: "The issues are clear cut--the opposing forces are grouping and sharpening their weapons. What is the citizen's role? It is simply this. He must stand up, be counted, and shout in a loud, clear voice-- I demand pure air." In spite of such democratic rhetoric, political participation is drastically limited. Even during presidential campaigns, most citizens do little more than vote--and millions do not even take that opportunity to participate. The literature on political participation indicates that most U.S. citizens are not well informed, not deeply involved, and not particularly active in political matters.[13] If defined as talking to friends and neighbors, writing letters, signing petitions, making complaints, advocating reforms, joining groups, and the like, political involvement for the control of air pollution is also limited to small numbers of people. Nonetheless, one useful measure would be a comparison of personal political activity on air pollution with other environmental and social problems. It could indicate that the participation rate is noticeably higher on air pollution matters than on most other problems in the specific air-quality region. And there is evidence to suggest that political interest in air pollution control is increasing, both in individual actions and in protest and professional group activities.

It would be a misinterpretation to equate participation rates with political support, for most people do not know what can or should be done to control air pollution. As Gilbert Seldes (1963: 346), pointedly explained to the 1962 National Conference on Air Pollution: "Everyone agrees on the facts--everyone says the situation is intolerable. And with the exception of yourselves and a few friends, everyone seems to feel that nothing can be done about it." In all available studies, this feeling of political impotence is clear. People do not participate because they do not believe that political alternatives are available or that political involvement could make a difference. To illustrate, let us quote the conclusions from an exploratory survey that we conducted with 45 residents of San Bernardino, California:

> Air pollution is not a politicized problem. No policy differences on control are seen in California between the two parties or their gubernatorial candidates. No one knows what the legislators can or should do about the control of air pollution. Almost no one is aware of any public agency involved in the control of air pollution--much less what their activities are. No policy proposals or possible control steps could be specifically cited, except that of smog devices for cars. No specific channels are viewed as open either for complaints or personal action. In substance, most people do not know how or even why they should become participants in public policy decisions on air pollution control [Loveridge, 1969: 5].

Policy stand as a concept draws attention to the policy preferences of the public and, more generally, to the question of political support. In a well-known definition, V. O. Key (1961: 14), said that public opinion "may be taken to mean those opinions held by private persons that governments find it prudent to heed." Politicians and specialists continually worry about what opinions to heed in making policy decisions. Most policymakers in a democracy think it desirable to make resource decisions in ways approved by the public. And, more important, they find themselves presented by public opinion, although uncertain tentative, and complex in effect, with some policy directions, limits, and sanctions.[14] The compelling question, however, is: What public opinions enter the air-quality policy arena. Or, in other words, how do policymakers know what the public endorses, prefers, or will support? The answer is: They don't, except in a gross, inexact, and perhaps erroneous manner.

There are now no reliable guides to the public's policy views on the control of air pollution. Three ways, all suspect, provide some information. Officials can proceed by hunch, intuition, and impression to estimate and interpret public opinion. Yet, these highly subjective views are open to error and subject to argument. Or, second, officials can depend on opinions expressed by groups. Unfortunately, most people have no organized channels in which to communicate their policy opinions. Citizen-protest groups are few in number, modest in organization, ill informed on alternatives, and without the expertise or resources to counter special interests that can make their positions known with vigor and persuasiveness. Citizens have few advantages in clarifying and stressing their views as against the lobbying of polluters.

And, third, officials can consult the results of public-opinion polls--although we do not know of a single instance where a poll has been commissioned to assist in the setting of air-quality standards. Most newspaper and commercial polls on pollution do not probe deeply into public attitudes; instead, they tabulate marginal distributions to superficial awareness or policy-choice questions. Anyone familiar with the limitations of polling data has to be dissatisfied with the incomplete and partial character of these results. Either/or responses on a few questions, except in the most general sense, do not tell officials much about basic public preferences or willingness to support control measures.[15]

The public's interest and preferences for environmental quality, and, specifically, the control of air pollution, are policy stands that have not been effectively communicated to policy councils. The task of measurement is not, however, a simple one. The man-in-the-street does not have the information, time, expertise, in short, the ability to develop well thought-out, consistent, or detailed policy positions on how to control air pollution. Policy innovation must by necessity be the work of specialists. The public nonetheless can react to major policy programs, prospective or actual, and indicate whether they support the cost, method, and objective. The prevailing assumption of many politicians and specialists is that, of course the public wants clean air, but they are unwilling to pay for it.[16] This supposition needs to be closely studied, especially as applied to the indirect costs of controls for stationary sources. Moreover, in his first national survey on pollution, George Gallup concluded:

> People are never eager to pay additional taxes, so these figures must be regarded as very encouraging. When these percentages are projected to the national adult population, it is clear that large potential funds are available to improve our environment [quoted by Cahn, 1969].

It is beyond our task here to develop specific measures of support and also depends on the context and purpose of the survey. Nevertheless, a survey of personal responses should devote a number of questions to direction, intensity, and latency of support for air pollution control policies.

As to doubts about the policy values of the surveys, any one of three possible contributions would warrant whatever expense and effort are involved. First, certain kinds of personal responses, as previewed earlier, could provide criteria for the formulation of air-quality standards. Second, survey results would require that politicians and specialists review the impact of pollution on citizens in addition to evaluating the costs of control on "progress" and economic interests. There has been an eclipse of citizenship to the point that the public's priorities are seldom voiced and become, in effect, marginal to control decisions.[17] Survey results could provide a new calculus for the weighting of control goals, strategies, and actions. And, third, pollution surveys should increase the pressure to adopt and enforce effective control programs, for they will register an increasing public consensus on the need for pollution control.[18] If counted, public concerns and preferences could provide major political incentives and sanctions in the setting of air-quality standards. Moreover, there is no reason for those who draft air-quality control programs to make impressionistic guesses on what the public wants or will support.

Interest Groups

Aside from the public, we initially thought that interest groups also required investigation. Yet, the question of "why" ultimately led to a decision to exclude interest groups. An inventory of the groups involved in air pollution politics could be taken, focusing on their internal characteristics--including values, structure and activities, and sources of influence--and detailing how they go about trying to achieve their aims, targets of their influence, their impact on pollution policies, and so forth. However, such an inventory, while perhaps important to an understanding of the politics of air pollution, does little to contribute directly to the control of air pollution.[19]

Policy Legitimation

After policies have been formulated, they must be legitimated, that is, approved by a legislative body. Air pollution provides a policy focus around which many legislative questions can be asked; however, few of these questions would seem to have any action payoff. One could write a detailed history of air pollution legislation, discussing what measures have passed (or not passed), and evaluating what future legislation is needed. Or a study could be undertaken to examine the clientele (party, governor, bureaucracy, pressure groups, constituency), specialized (legislative leadership, subject-matter experts, staff help), and incidental (personal characteristics, group membership) influences on air pollution legislation. Or the procedures could be studied by which air pollution bills become law in order to see who gets what, when, and how (stages would include committee hearings, floor debates, conference committees). Or rollcall analyses could be conducted of votes on air pollution bills, and so on. Nevertheless, despite the relevance of these studies to the tools and past research of political scientists, we do not find that such efforts could be usefully translated into "practical solutions."

Policy Implementation

While legislatures adopt control measures, these policies must be implemented by administrative control agencies. "The application of policies," explained Dan Nimmo and Thomas Ungs (1967: 402), "is the task of bureaucracy; it is one which shifts the resolution of interest conflict from the legislature to the administrative arm of government." The pivotal feature of air pollution control is the compliance process where emission standards are enforced. Three possible answers exist as to why air pollution remains a serious problem. First, the state of the technology cannot deal with the complexity of control problems; or, second, the commitment of the polity is in no sense decisive and in effect postpones resolution; and/or third, the legislation is satisfactory, however, its implementation is uncertain and largely ineffective. While all three answers are probably partially correct, the third--enforcement of emission standards--deserves special scrutiny as a possible explanation.

Political scientists have increasingly emphasized the preeminent role that administrative agencies play in U.S. politics. Administrative officials are in a commanding position to influence public policy because they combine expertise and a near monopoly of technical information with an ability to structure public attitudes and preferences. Nevertheless, while possessing major resources, repeated studies have documented that administrative agencies often become a reflection of the system of group pressures. And, in the case of air pollution control programs, the influence of administrative agencies depends especially on the good will and support of the groups they are set up to control. (Unfortunately, most pollution-enforcement decisions are invisible to and too complicated for the public-at-large.) Briefly, the stakes of the pollution game can, in many ways, be won or lost in the politics of administrative-control agencies.

In California, the problems of compliance have to be divided into those for moving and stationary sources. The State Air Resources Board sets and to some extent enforces emission standards for moving sources. The problems of compliance, although difficult to resolve, are well established. No cars--except for a few assembly-line prototypes--are tested to see if they meet the emission standards of the State of California. However, legislation requires, beginning in 1973, that every car sold in California meet specified emission standards. But proposals for annual inspection face objections of costs versus benefits and, more importantly, the lack of monitoring devices that are low in cost, easy to use, and still accurate. The problems of compliance for vehicular sources can, in short, be clearly identified and do not require the expert advice or research of political scientists.

In contrast, stationary sources present important compliance problems on which we have little information. Within a broad state mandate, county control districts can adopt and enforce their own stationary standards. Past research would suggest that these agencies are particularly open to the influence of clientele groups.[20] If these agencies determine what standards are adopted and, more importantly, how they are enforced, what differences exist in the characteristics and performance among the various county control districts? And, more specifically, how effective are they? For example, John Maga (1965: 27), wrote: "Bay Area enforcement efforts are less effective than those in Los Angeles." Moreover, the counties vary in their per-capita expenditures for pollution control from less than one cent to sixty cents in Los Angeles. (One source cited forty cents per capita as the minimum for an effective program--only San Bernardino and Los Angeles meet that standard.)[21] The implementation approaches and results of county air pollution control agencies merit careful evaluative research.

The performance of administrative-control agencies is largely determined by two classes of factors, organizational and sociopolitical. Organizational factors include the administrative structure, enabling legislation, enforcement methods and procedures, staffing, and especially the personal characteristics of the chief administrative officer, size of budget, financial and technical assistance from state and federal agencies, approach to community relations, and so forth. These organizational factors, among others, should significantly influence what standards are set and enforced.

Beside organizational characteristics, the influence of sociopolitical factors on the stationary control process cannot be overlooked. The economic forces for growth and progress place strong pressures on control agencies to be "judicious" in implementing air-quality objectives. What discretion and strength county control agencies have depends to a large degree on the political support they receive from their supervisors, administrative office, public-at-large, and perhaps most importantly, clientele groups. Without strong political support, control agencies face the prospect of offering the symbols of control but, in fact, allowing the control process to be loose and "flexible." Therefore, the call is for a rigorous study of the performance of county control agencies, with particular attention to their organization and political environment.

Despite the fact that state and county control agencies can substantially determine the kind of air we breathe, careful studies of their performance have not been conducted. We do not know of even one systematic study of an air pollution control district, much less the seventeen or more in California. At the national level, the task-force report on air pollution by the Ralph Nader group headed by John Esposito (1970), is a partial exception. The task force intended to examine the operations of NAPCA but quickly found it necessary to expand its focus to include the forces that operate on the agency from the outside--from business, the Congress, and the public. Although descriptive in approach, unclean in procedures, and sometimes over-dramatized in its conclusions, the report nevertheless presents an important analysis of the constraints and weaknesses of NAPCA. Ralph Nader stated in the Introduction that "its most significant contribution is its analysis of the collapse of the federal air pollution effort starting with Senator Edmund Muskie and continuing to the pathetic abatement efforts and auto policies of NAPCA." Beyond this contribution, the report attempts (and, we think, successfully), to illustrate the relation between clientele influence and standard setting and enforcement at the national level.[22] Yet, it should be emphasized that the report is not nor does it pretend to be a systematic study of the performance of NAPCA.

While government intervention is increasing in social and economic activities, few studies of the behavior of administrative agencies exist. And studies that try to investigate the efficiency or effectiveness of administrative agencies are especially rare.[23] Political scientists, however, are beginning to investigate the outputs of administrative agencies. One excellent example is the work by James Q. Wilson (1968), on the behavior of police departments; eight communities were included. However, some twenty-five cities were initially studied, and reports were written on each for purposes of comparison. While the study did not collect sufficient numerical data to allow for rigorous analysis, it does represent an important attempt to compare the behavior of similar administrative agencies. Wilson (pp. 32-33) has explained his rationale as follows: "I feel political science has, in some degree, a peculiar mission and competence: to think simultaneously about the quality of the ends that are served and reasons why those ends, and not others, are in fact served." A new research thrust of political science promises to be the study of policy outcomes and their relation to the behavior of administrative agencies.

Beside describing and explaining the performance of administrative agencies, evaluative research also leads to recommendations for possible changes. After studying how agencies in fact behave, political scientists should find themselves in a position to offer organizational prescriptions and political alternatives. The administrative agencies expected to set and enforce emission standards should be sophisticated and viable instruments, possessing what initiative and powers are necessary to achieve the air-quality goals defined by the polity. Without sound evaluative research, it is difficult to assess the successes and failures of the implementation process and, in turn, to recommend new incentives and controls.[24]

Research on implementation should, however, extend beyond the present performance of administrative-control agencies. Many scientists see little hope for clean air, and other scientists see even less for a tolerable urban future unless drastic, concerted, perhaps radical policy programs are adopted. For example, University of California ecologist, Kenneth Watt, offered this prediction of the fate of Los Angeles:

> Projections in the Archives of Environmental Health show a potential smog disaster hitting the Long Beach area during the Winter of 1975-76. Thousands of people may die. I'm now busy trying to figure out where other potential smog disasters will hit. The incredible thing is that nearly all scientific and medical evidence necessary to predict the end of Los Angeles is right here in the university library [quoted by Rapoport, 1970: 84].

And a similar prognosis for the Bay Area was made by Stanford pharmacologist, Robert Dreisbach (1970), who said: "Ecological catastrophe is here now and we only need to open our eyes to see it." He contends that the Bay Area is burying itself in its own garbage and choking to death from uncontrolled pollution.[25] If these forecasts are even partially accurate, new policy programs are required that are more than incremental improvements in the status quo. Projected increases in population, moving vehicles, miles traveled, new kinds of pollutants, power requirements, and growth of industry would seem to call for new, major controls and thus new, institutional forms and incentives. Control officials, scientists, and planners frequently cite the urgency of reform, yet are either unfamiliar with or pessimistic about the possibilities of policy changes. Political scientists can offer their expertise in evaluating and designing new institutions and, perhaps even more important, proposing possible strategies to reach avowed policy objectives. Future scenarios for control implementation call for political imagination as well as sober reflections on political feasibility, for without the two in combination, the prospects for clean, even tolerable air appear unfortunately bleak.

RESEARCH PROSPECTS AND PROJECTS

An assessment and review of the literature in political science poses two research questions: What do we know? And what can we do? Four kinds of projects merit attention: (a) responses, support, and demands of the public; (b) effectiveness of administrative control agencies; (c) analysis and evaluation of policy strategies and future control alternatives; and (d) a general study of the politics of air pollution across the nation. The projects represent possible

POLITICAL SCIENCE AND AIR POLLUTION

contributions to efforts to control air pollution. Moreover, the projects are viewed as feasible and within the research expertise of political scientists. In other words, the selection emphasis is on applied research projects that have some potential payoff in terms of the policy process.

Views of the Public Toward Air Pollution and its Control

The responses, support, and demands of the public can be studied by survey research and, to a lesser extent, by more intensive interviews with specific subgroups and perhaps in experimental situations. Data on public attitudes toward air pollution should have several noteworthy consequences. First, in an important sense, democratic rhetoric and policy practice can be consummated, for the responses and demands of the public can be placed on the official docket and into the process of policy-control development. Second, attitude studies can identify and emphasize major personal effects of air pollution that are presently not evaluated. And, third, attitude studies should identify the extent to which the public demands and will support pollution-control measures and, in turn, what information, guidance, incentives, and sanctions the states and the federal government must present in order for its programs to be understood and supported by the public.

Effectiveness of Air Pollution Control Agencies

There is a dearth of research and analysis on the effectiveness of administrative-control agencies. The literature of air pollution control is replete with engineering and science applied to technical problem solving. Relatively unexamined, however, are the administrative agencies that apply the highly developed technology of control. Yet, many of the major control decisions are made within these administrative agencies. Too often, the implementation process tends to provide reassuring symbols for the public and concrete benefits for organized interests. However, the performance of control agencies remains unstudied: What are the effects of different organizational characteristics and sociopolitical conditions? Research directions are twofold. First, criteria should be developed to measure agency effectiveness, for until such criteria are framed, it is difficult to study and evaluate agency performance. And, second, field data should be collected from state and local major control agencies that focuses on characteristics, conditions, and performance.

Policy Strategies and Control Alternatives

Two kinds of research projects are called for in the area of policy strategies and control alternatives. The first would take up alternatives to present control methods and would also examine as well as evaluate their probable consequences. In contrast, the second would explore and analyze the larger questions to determine what institutions, incentives, and controls will be necessary for the future development and habitability of the major air basins and what political steps can and have to be taken to reach desired objectives. The first project would be based primarily on specific research work and would deal with such questions as competing concepts of control, regulatory processes, systems of incentives and punishments, and problems of representation in policy development.26 The second project would be interdisciplinary in approach and would focus on future pollution prospects and what must be done if tolerable air is to be maintained. The specific contribution of political scientists would be to examine the political steps required to meet the objectives set by planners and scientists. Plans and reforms are of little value unless they can be translated into policy outputs and implemented. Increasing evidence suggests that we need to reevaluate the structures and objectives of our society in terms of what will be necessary for environmental survival.

Handbook on the Politics of Air Pollution

There is no available handbook on the politics of air pollution. Citizens do not have access to any source that can detail the extent and nature of the problem, the legislation passed, the major actions taken, or the policy problems of control. A study of the politics of air pollution, self-consciously designed for the interested layman, would be a valuable public service. As a project, this study would not command the priority of the first three, yet for purposes of intelligent citizen action and support, it would seem important to clarify the confusion surrounding air pollution. For consistent and meaningful public participation, citizens should be informed as to what is the problem and what has been done and <u>should</u> be done to achieve clean air. Otherwise, the issues of air pollution control become lost in the complexities of pollution, the rhetoric of politicians, and the myopia of "ecotactics."

NOTES

1. In the Presidential Address to the Sixty-fifth Annual Meeting of the American Political Science Association, New York, September 2-6, 1969, David Easton offered perhaps the best explanation for the emergence and characteristics of postbehavioralism. See Easton (1969).

POLITICAL SCIENCE AND AIR POLLUTION   81

   2. An excellent discussion of the relationship between political science and public policy can be found in a collection of papers edited by Ranney (1968). For a more general introduction to policy analysis, Lindblom (1968) and/or Mitchell and Mitchell (1969).

   3. See, for example, California's <u>Environmental Quality Study Council: Progress Report</u> (1970). If adopted and implemented, the recommendations for land use and, to a lesser extent, air quality, would in effect significantly resolve the problem of air pollution in California. Yet, prospects for approval seemed remote, for many of the recommendations clashed with strong, vested interests in the status quo.

   4. See Loveridge (Paper, 1970: 2-28); while the selected text from the paper has not been significantly changed, we have modified the footnotes to fit the note sequence of this chapter.

   5. This chapter will draw on many of the basic concepts and established findings in political science. Though perhaps not cited as such, our observations on environmental politics are frequently extrapolations from past research work on the policy-process problems.

   6. The results of interest-group bargaining are well illustrated in an article by Stein (1969: 34); she wrote that the supersonic transport (SST), had high priority and abundant funding, although it would serve about 5 percent of the population, while the nonpolluting-car program had no priority and virtually no funds, although it would benefit almost everyone in the country.

> Although noise pollution from the inescapable sonic booms could create the most widespread environmental blight in the nation's history, the U.S. government has already allocated more than half a billion dollars for the SST program and may end up spending between $1.3 and $4.5 billion to see the venture through. . . a different kind of environmental blight--air pollution--continues to grow in scope and intensity. America's 97 million motor vehicles. . . are producing more than 60 percent of that pollution. Yet the Senate has not even acted on legislation to spend a comparatively mere $3.5 million toward the devleopment of automobiles that do not pollute the air.

The reasons for such resource allocation can largely be explained by the respective influences and interests of the aircraft and automobile industries. If committed to the process of political bargaining, environmental issues will usually be won in legislative battle by the corporate interests and lost by the public interests.

7. In a newspaper column, Art Buchwald (1970), hit this point hard in what he called the problem-of-the-year contest:

> Each year the American people, with the help of the news media, decide which Problem they will be the most concerned with for the next twelve months. Last year, if you remember, it was "Crime in the Streets." The year before that it was the "Wars." Past winners have included "hunger," "poverty," "desegregation," and "Cuba." . . . Well, I'm going over now to pick up the white envelope which will tell us what Problem will reign supreme in 1970. . . . I tear off the top and . . . The Problem of Year is Pollution! Miss Pollution is the new Queen. Let's hear it for Pollution.

8. The new, federal, Environmental Protection Agency (EPA), may be an exception. In particular, the agency seems to have the platform, support, and powers necessary to translate federal good intentions into more effective pollution-control policies. For example, on January 29, 1971, EPA proposed--and was to set in motion procedures to enforce--nationwide air-quality standards that were much tougher than past federal air pollution laws. For background and organization of EPA, see "Man's Control of the Environment" (1970), or <u>First Annual Report of the Council on Environmental Quality</u> (1970).

9. Conine (1970), wrote:

> The impression grows, however, that the great mass of politicians, students, and assorted other environmental crusaders have given little thought to the cruel and perplexing paradox which we face. Namely, that in fighting to save humanity as a whole from slow poisoning or strangulation, we could find ourselves denying to millions of human beings the means of escaping poverty.

10. Russel Train (1970), Chairman of new, federal, Council on Environmental Quality, note:

> Even were we to eliminate all forms of environmental pollution, we would still not have guaranteed a high-quality environment. Environmental quality is a far more complex, more subtle objective. It involves the development of new attitudes and new values. Thus, while we must make the investments and achieve the technological breakthroughs necessary to clean up our environment, we must at the same time develop a new perception of man's relation to nature, learn to control

our own numbers, develop effective land-use policies, and find new measures of public and private success which emphasize quality rather than mere quantity.

11. For a detailed examination of the legislative process, see Bailey, (1950); Clapp (1963); Fenno (1967); Froman (1967); and/or Keefe and Ogul (1968).

12. See Loveridge ("Types, Ranges," 1970); we have included only selected parts of the text, and, as above, we have modified its footnotes to fit the present sequence.

13. See, for example, Lane (1959) or Milbrath (1965).

14. For example, R. Kovitz (1967: 27), wrote: "After six years of Board activity, the Motor Vehicle Pollution Control Board (MVPCB), realizes that it can go no faster than public acceptance of its program."

15. See Verba, et al. (1967).

16. By way of an illustration, Kovitz (1967: 26) declared: "Every Californian is against smog. He is more than willing to pass laws that will get rid of it. The problem arises when implementation of those laws costs him money."

17. See Robert Pranger (1968: 39-86).

18. Most political observers agree that there is an increasing public consensus on the concept of clean air as a necessary requirement for improving the quality of our total environment. An editorial in a conservation magazine, "Mr. Hickel Has a Choice" (1969), stated directly the position we find to represent the developing air pollution stance of the U.S. people:

> We question why the richest nation on earth should be anything but publicly forthright and aggressive about stopping the national disgrace of pollution NOW. The country should not wait until the impact on environment is worse, nor until the problems have been studied to death, nor until the last buck has been made from causing the pollution itself.

Yet, despite public anguish and anger, the problem of air pollution--according to many sources--is getting worse. Pollution surveys should provide a political clout for new laws and new emphasis on pollution controls.

19. Why some advocacy groups are effective and others are not is, no doubt, an important research question. For an extended discussion on the possible importance of advocacy group in air pollution research, see Chapter 2. Nevertheless, in terms of probable control results, we could identify no interest-group study that has had an applied pay-off for a major policy question.

20. Past research refers here to over sixty student papers completed from 1966 to 1970 at the Riverside campus of the University of California. Students studied every control district in Southern California, investigated the activities and policies of a number of stationary polluters, conducted extensive interviewing for background and policy studies, and reviewed and evaluated what literature was available on the process of air pollution regulation and enforcement.

21. Following is a statement from Air Pollution Control (1967:3):

> During the spring of 1966, National Association of Counties Research Foundation personnel made field visits to local air pollution programs of all sizes across the United States. In addition, a detailed questionnaire was distributed to more than 300 local programs. Based upon in-depth analysis of the effectiveness of these existing programs, this foundation recommends an absolute minimum expenditure of 40 cents per capita for all local regulatory programs.

22. The Nader task force, for example, states in the conclusion to their report:

> Throughout this book, the Task Force has illustrated how the public's hope for clean air has been frustrated by corporate deceit and collusion, by the exercise of undue influence with government officials, by secrecy and the suppression of technology, by the use of dilatory legal maneuvers, by special government concessions, by high-powered lobbying in Congress and administrative agencies and--in ultimate contempt for the people--by turning a deaf ear to pleas for responsible corporate citizenship. [Esposito, Vanishing air, 1970: 299].

23. While specific studies of administrative agencies are sparse, an extensive literature has developed on such subjects as the administrative process, regulatory process, and policy and politics; see, for example, works cited in the Bibliography.

24. Sheldon Samuels (1970:23-24), Chief of Field Services,

NAPCA Office of Education and Information, concluded an evaluation of the contribution of social-science research to air pollution control as follows:

> The behavioral research described here has provided substance for the ordering of agency priorities and for the basis of public decision. Basically, the results were not contributions to science but to the evaluation of programs. In a broad sense, the author feels that the evaluative role is the only pragmatic one for <u>applied</u> behavioral science in the actual administration of government programs. Namely, it is most practical to be concerned less with what will work and more with what <u>does</u> work.

Daniel Moynihan (1967:9), makes the same point when he recommends that Congress "should now establish an Office of Legislative Evaluation in the GAO. . . . This office would be staffed by professional social scientists."

25. Specific quotes are taken from a news conference at Stanford reported by Marshall Schwartz in the <u>San Francisco Chronicle</u> November 7, 1969, Sec. I, pp. 1, 28. For documentation leading to the pessimistic prognosis of the future of the Bay Area, see Robert Dreisbach (1970).

26. See Michael Reagan (1970).

CHAPTER

5

**AIR POLLUTION
AND LEGAL
INSTITUTIONS:
AN OVERVIEW**
James E. Krier

INTRODUCTION

Even the staunchest of government hands-off conservatives agree that some type of government intervention is generally called for in the case of air pollution. Most environmental problems, this one among them, must be solved by government action if they are to be solved at all. This is so simply because there are no viable alternative mechanisms to government control to restrain air pollution and other environment-debilitating activities. (One could argue, of course, that both public opinion and a sense of responsibility work as restraints on socially undesirable conduct, but in the case of pollution, these usually cannot be considered "viable alternative mechanisms" to some form of governmental restraint; the reasons for this conclusion are suggested at various points in this chapter.) Without the imposition of governmental restraints, firms and individuals could pollute with utter abandon, because imperfections in the market mechanism would, in most cases, permit them to do so without taking account of the costs imposed on others by their polluting activities.

When we speak of governmental restraints or governmental intervention, we refer to restraints on or intervention in regard to a specific problem, such as pollution. In a pervasive way, government always intervenes in social affairs by holding a monopoly on the legitimate use of force. But for this monopoly, pollution could dissapear even without

---

A revised version of this paper appeared in Volume 18 of the U.C.L.A. Review.

specific government intervention. One unhappy person could, for example simply blow up all the polluters without fear of government reprisals. (Unhappy people occasionally act in spite of possible governmental reprisal. Consider, for example, the activities of "The Fox," "sort of an antipollution Zorro, who has been harassing various companies" to make them cut down on polluting activities. The Fox has been doing such things as plugging up the drainage systems and sealing the chimneys of firms in the Chicago area. " 'Nothing seemed to make them stop. So I decided that even if I was only one man, I'd do something. I don't believe in hurting people or in destroying things, but I do believe in stopping things that are hurting our environment,' " the Fox said.[1]

When used as a garbage dump, the air becomes a factor of production because, once used this way, it is less suitable for other uses. Like other factors (such as raw materials and labor), it should be purchased, and its "cost" should ultimately be reflected in the price the manufacturer charges for his product. But the market cannot function by itself to bring about this result in the case of air pollution. The market mechanism alone cannot force the polluter to pay for the right to use the air as a dump because there is no way the market can withhold air from the polluter; the air does not come in marketable packages. All of the receptors of pollution, in theory, could get together and pay polluters to stop, but as a practical matter this will never come about. In the first place, each receptor would be prone to take the position that he would be a fool to pay because the payments made by others would benefit him whether he paid or not. No person who paid a polluter could exclude these free-riders from the benefits of his payment. As a result, those few who might be willing to make payments to polluters would find it necessary to spend enormous sums, more or less equivalent to the sum a polluter would demand if all receptors were paying.

Moreover, payments would have to be made to all polluters. A payment to one would obligate it to change its ways, but it would have no bearing on the conduct of others; they could continue to exploit the same air. (Indeed, it might encourage the appearance of new polluters who would demand similar payments in exchange for giving up the power to pollute.) Because the market would not function to charge for the right to pollute or the right to breathe, there would be no prices paid or payments foregone by pollution manufacturers that would convey to them information about the social utility of their conduct or give them an incentive to change their ways (Dales, 1968; Goldman, 1967; Ruff, 1970.) If quality improvements are to be realized, government intervention is called for (Mills, in Walozin, 1966). One must always consider, of course, whether the benefits realized through quality improvements are worth the costs (of government intervention and of opportunities foregone), involved in achieving them, a point to which we shall return.

For those concerned with the form and function of legal institutions, there is little comfort in the advice that government must take a hand in solving the air-pollution problem. Because government can only intervene through its agencies (through its executive, administrative, legislative, and judicial bodies), important questions remain, including these: Which government or governments--federal, state, local or regional--should intervene, and in each case, through which of its institutions, and when, and in what manner? How has government intervened in the past, and with what success or failure? What indications are there of needs for change; what direction should change take; and what risk is there that any change will, in solving the perceived problem, create other problems of greater dimensions?

This chapter assays the state of our knowledge about questions such as these and, based on that assay, suggests some directions for fruitful research. The first section of the discussion sets the chapter's major theme. It outlines a conceptual view of the pollution problem that is particularly meaningful to those concerned with improving and developing institutions to cope with air pollution. As the discussion indicates, the problem of allocating the air resource to competing uses (such as breathing and polluting) is both an empirical one (some of the issues that arise can probably be settled with data which is now available or which can be generated), and a political one (as to some of the issues, it is unlikely that we will ever have the necessary data, and we must proceed on the basis of hopefully democratic decisions).

Discussion in the first section gives special attention to the political nature of the problem and considers the implications for those concerned with legal process. The second section discusses the pollution problem from the standpoint of judicial control. It suggests some of the objectives that pollution litigation should serve and considers the extent to which those objectives have or have not been realized. In this part of the discussion, the ways in which the courts have gone about defining property rights in the air resource is given some attention. Finally, this section isolates two classes of shortcomings suffered by judicial process with regard to pollution litigation-- problems of proof and problems inherent in the judicial process--and suggests some ways in which these shortcomings might be overcome. The third section looks at pollution-control legislation and its administration. More particularly, it compares programs of regulation, subsidization, and pricing; reviews the strengths and weaknesses of each; and suggests some research strategies that could very likely lead to constructive change.

LEGAL PERSPECTIVES

As a beginning, it may be worthwhile to return to a consideration of the nature of the air pollution problem both to examine it in greater detail and to consider the role of political processes in the function of problem definition.

The Nature of the Problem

Lawyers and legal scholars have been far too ready to accept shallow definitions of the cause of air pollution. They have too often seen the essence of the problem as the presence of gases and particulates in the air or as their escape from smokestacks and other sources (See, e.g., Cowan, 1955; Note (a), 1968). Of course, these are causes of the problem at one level of analysis. In another sense, however, they are only symptoms of underlying causes, the results of patterns of human behavior. While this is all to some extent a matter of definition, it is not a matter to be settled arbitrarily, for one's perspective affects the action he takes. The smokestack viewed as a cause calls for an attack on the smokestack--put a control on it. This is very often the response of the engineer or technologist to the problem of pollution. The difficulty with this view is that it overlooks the possibility of attacking the underlying cause, of changing the incentives that lead to smoke production.

However, not all legal scholars have taken the perspective described above. Since the late 1960's, a good deal of legal research that discusses the sorts of market breakdowns which produce environmental problems has appeared (Baxter, 1968; Atwood, 1969; Wright, 1969; Hagevik, 1968; and Delogu, 1969). And such discussion has led to valuable suggestions for dealing with social-cost problems akin to the air pollution problem. For example, with regard to the sonic boom, Baxter (1968), concluded that the external costs imposed by the SST and its ear-cracking boom could be efficiently internalized (i.e., imposed on the SST), through the creation of a statutory fund supported by contributions from SST operators. This fund would be subject to damage suits, and strict liability would be imposed on it; moreover, social costs estimated by the government not to have been recovered in the damage suits would be assessed against the fund and made payable to the U.S. Treasury (Baxter, 1968: 53-57).

Baxter's analysis is interesting both because it illustrates the creative and constructive institutional mechanisms that can be built upon an understanding of social problems and because it is based on a view of the social-cost problem with which some scholars have

AIR POLLUTION AND LEGAL INSTITUTIONS

taken issue. Baxter's view of the problem of social cost is plainly reflected in the following:

> Whenever a business activity imposes costs for which it is not required to pay, or bestows a benefit for which it is unable to charge, the price mechanism will not ensure a proper allocation of resources, even in a world of perfect competition.
>
> One of the primary functions of a legal system should be, then, to ensure that costs such as these do not remain external to the enterprise that creates them [Baxter, 1968: 39].

But, which enterprise "creates" the social cost? Baxter's answer is consistent with the Pigovian tradition that emphasizes the divergence between social and private cost. The private costs of the SST are less than the uncompensated costs its boom imposes on society, and unless that difference in cost is brought to bear on SST operators, resource misallocation will most likely result: Too many SSTs will fly, and too many booms will crack. Similarly, in the case of the factory that produces smoke, this analysis would suggest that, because there is a divergence between the private and social product of the factory, it would be desirable from the standpoint of resource allocation to make the factory liable for smoke damage or to impose on it a tax equal to the damage it causes or, finally, to exclude it from areas in which the smoke is likely to cause harm to others (Coase, 1960).

Ronald Coase, an economist, took issue with both this analysis and the conclusions it suggests. He argued:

> The traditional approach tended to obscure the nature of the choice that has to be made. The question is commonly thought of as one in which A inflicts harm on B and what has to be decided is: How should we restrain A? But this is wrong. We are dealing with a problem of a reciprocal nature. To avoid the harm to B would inflict harm on A. The real question that has to be decided is: Should A be allowed to harm B or should B be allowed to harm A? The problem is to avoid the more serious harm [Coase, 196 1960: 2].[2]

Coase demonstrated that a pricing system which operated without cost would automatically produce results which avoided the more serious harm, i.e., results which reflected optimal resource allocation. These results would follow, if the pricing system operated without costs, whether or not polluters were liable for the damages they caused (Coase, 1960: 2-15). In other words, in a world with a frictionless market, liability rules are neutral from the standpoint of resource allocation.

A simple example will clarify this point (which is so contrary to most of the assumptions of law-trained people). Assume that a house on a plot of land is let by its owner at an annual rent of $1,000. A factory buys an adjacent piece of land and begins production activity that sends sooty and unpleasant-smelling smoke over and into the house during most of the year. Suppose that the factory makes a profit of $2,000 per year as conditions stand, but to abate its pollution (with control equipment, or by changing fuels or production methods or by moving to a different area) would reduce the profit to $1,500. Suppose too, that under the changed conditions the landlord can obtain only $800 in rent, and that putting his property to a different use or abandoning it altogether would leave him even worse off.

In a system in which the factory's activities do not give the landlord a right of action for damages, the landlord would bear the annual rent loss of $200, and the factory would continue to pollute. The landlord might offer to pay the factory up to $200 to abate its pollution. That offer, however, would be refused by the factory because abatement would cost $500. The landlord would not rationally put his property to a different use or abandon it, for that would leave him even worse off than settling for the decreased rent payments.

Suppose, however, that the law makes the factory liable for damages to the landlord. The damages in this case would be $200 in loss of rent and a rational factory owner would pay that amount, as any other means of avoiding the liability would cost him at least an additional $300. Presumably, landlord (and tenant) and a factory carrying on operations as usual would have been satisfied. And, just as with the case where the factory was not liable, the same amount of pollution would exist. There would be a different distribution of wealth between the landlord and the factory owner under one scheme than the other, but resource allocation would be the same in either case. (The question of wealth distribution is by no means unimportant; see the discussion _infra_.)

Of course, liability rules are neutral only in a world that does not exist--a world where people are rational; where they have or can obtain without the cost of time, effort, or money all the information they need to exercise their rationality; and where they can freely bargain without any effort or expense (other than the consideration exchanged in the bargaining process), with those whose activities bear on their interests. Obviously, these suppositions that underlie the assumption of costless market transactions make that assumption "a very unrealistic" one as Coase (1960: 15) points out.[3]

AIR POLLUTION AND LEGAL INSTITUTIONS

> In order to carry out a market transaction it is necessary to discover who it is that one wishes to deal with, to inform people that one wishes to deal and on what terms, to conduct negotiations leading up to a bargain, to draw up the contract, to undertake the inspection needed to make sure that the terms of the contract are being observed, and so on. These operations are extremely costly, sufficiently costly at any rate to prevent many transactions that would be carried out in a world in which the pricing system worked without cost.

When all these information and transaction costs are taken into account, the sorts of bargains or market transactions that would efficiently allocate resources are unlikely to come about, for the costs of the bargaining process might well outweigh the benefits of the bargain. The landlord and factory owner in our illustration, for example, might not bargain if that process required substantial legal fees.

Under these circumstances, the initial delimitation of rights and liabilities by the legal system does make a difference. When the costs of bargaining are greater than the benefits of the bargain,

> [The] granting of an injunction (or the knowledge that it would be granted) or the liability to pay damages may result in an activity being discontinued (or may prevent its being started) which would be undertaken if market transactions were costless. In these conditions the initial delimitation of legal rights does have an effect on the efficiency with which the economic system operates. One arrangement of rights may bring about a greater value of production than any other. But unless this is the arrangement of rights established by the legal system, the costs of reaching the same result by altering and combining rights through the market may be so great that this optimal arrangement by rights, and the greater value of production which it would bring, may never be achieved [Coase, 1960: 16].

The bearing of Coase's analysis on the form and function of legal institutions was outlined in a cogent essay by Calabresi ("Transaction Costs," 1968). The balance of this section will sketch Calabresi's ideas (in the above article) so that reference may be made to them in subsequent discussion.

Bear in mind that the central concern here is optimal allocation of the air resource to competing (polluting and nonpolluting) uses. Resource misallocation exists when a change in resource use would produce a situation where those who gained from the reallocation could fully compensate those who lost and, after this compensation process, some persons would be better off than before. Optimal allocation exists only when no reallocation could benefit someone at no necessary cost to anyone. The costless and rational bargaining described by Coase would obviously produce optimal-resource allocation, because a misallocation would exist whenever the situation could be improved by bargains and, with no costs attached to the process, rational people would always bargain under such circumstances to the optimal point. From this, it follows that all resource misallocations can be corrected by the market except insofar as information and transaction costs and other impediments (such as criminal sanctions which do not permit one to buy his way out) create bargaining obstacles. The resource-allocation goal is to produce, as closely and as cheaply as possible, that allocation which would result if the market worked costlessly (Calabresi, "Transaction Costs," 1968: 68-69).

The relevant issues then become: Is this best accomplished by government intervention, by reliance on the market, or by some combination of these? If government intervention is required to achieve the resource-allocation goal, what mix of institutional forms and means should be used? Government can intervene at the federal, the state, the local, and the regional level. At each of these levels, it can intervene through (or through any combination of) executive, legislative, judicial, and administrative bodies; moreover, each of these bodies can intervene in a variety of ways. The legislature, for example, can regulate, subsidize, tax, or simply gather and disseminate information. Each of these means probably brings with it different costs. Some means may also bring higher relative chances of reaching wrong results. The problem is to pick, largely by guesswork, the institutional response that is likely to bring the minimal mix of risk and cost and then to decide whether those risks and costs are worth the benefits which will hopefully flow from the reallocation of resources (Calabresi, "Transaction Costs," 1968: 69).

Of course, there is not a great amount of knowledge about what allocation of resources is optimal, nor is there much information in any rigorous sense about the benefits resulting from incremental steps toward that allocation (Calabresi, "Transaction Costs," 1968: 69-70). More than an occasional voice can be heard to cry for absolutely clean air at any cost whatsoever.[4] However, most sensible people have come to realize that there is a point where the benefits from a further percentage increase in air quality are outweighed by what must be given up elsewhere to achieve that increase (as well as a point where information bearing on further possible benefits and costs is not itself worth the expense of acquisition). As a practical matter, however, we shall never know, in anything approaching rigorous, quantified terms, just what that point is (Ayres, 1969; Kneese, 1967).

For one thing, there is no way to quantify all of the variables required by the calculus involved. Even if there were, the expense of reducing all of them to quantitative terms would probably be overwhelming. As a result, it is necessary to make collective guesses about whether the benefits of further increases in air quality are worth the costs of attaining them. Is action appropriate, or should the market be left to work things out as best it can?

In facing this question, two guidelines should be considered. The first is to respond in ways which, if wrong, can be most cheaply corrected by unassisted market action. This guideline gives us insights into whether government should intervene. (Could the market more cheaply correct an error resulting from some form of regulation or from liability rules?) The second guideline is to take actions and forms of action that serve other goals with, at worst, no predictably unfavorable impact on resource allocation. Government action (or government action of a particular sort), may produce a favored distribution of income, for example, with no discernible adverse effect on resource allocation (Calabresi, "Transaction Costs," 1968: 70). Indeed, with respect to the air pollution problem, the income distribution and resource-allocation goals often appear to be compatible.

All of the above remarks have great bearing on the sorts of institutional responses that should be made to the air pollution problem. Before considering them in that context, however, it may be worthwhile to discuss briefly their implications for the air-quality decision process.

The Problem and the Political Process

The air-quality decision necessarily depends on collectively made hunches about relative costs and benefits. These hunches are loaded with subjective value judgments (which undoubtedly explains why the central point of most arguments over environmental-quality questions is coming to be, at least implicitly and unconsciously, an analysis of costs and benefits at the margin). Consider, for example, the case of a big factory planned for a depressed city in the Deep South. The factory might foul the air for the people living there, but at least it would give them jobs. Most people sense that the issue is whether the new prosperity is worth the loss in good breathing; as one would expect, however, that issue tends to be solved differently by different classes of people involved in the decision. Low-income groups quite uniformly weight the gains and loses in such a way as to make the factory a worthwhile addition to the local economy; the more affluent use accounting methods and measures that yield the opposite result (Simon, 1970: 11). This divergence in viewpoint illustrates the nature of the air-quality decision.

The relevant inquiry then becomes typical: How should institutions of the political process be shaped to bring them most democratically to bear on these value questions of air quality? Consider the machinery of federal air pollution control legislation as it functioned before the 1970 amendments. With respect to fixed-source emissions, the law provided that HEW should develop air-quality criteria describing the nature of various pollutants and their effects at different levels. The states were then to translate these criteria into air-quality standards (ambient-air standards) and implementation programs (such as emission standards) designed to achieve the desired level of air quality.5 Where should the value judgements that are an essential ingredient in the air-quality decision have come to appear in this flow-chart of decisionmaking? More particularly, how should all the dirty and practical, but nevertheless unavoidable, considerations of politics have come to bear on the decision process? One would think not at the criteria stage:

> Air-quality criteria, expressed in terms of pollutant concentration and duration of exposure, describe relationships between air pollutants and effects on health and welfare. Such criteria summarize what is know about the effects of pollutants in the atmosphere to provide a realistic base for selecting air-quality standards ["Air quality Criteria for Sulphur Oxides," 1967: iv].

Thus, the criteria stage would at first appear to be a pure, and a purely scientific, phase of the decision process, based on objective and quantifiable standards, such as health considerations, damage to plants and animals, visibility, and metal corrosion ("Air Quality Criteria for Sulphur Oxides," 1967: vi-vii). There would appear at this stage to be no room for "political" or value considerations.

Yet, the "relationships between air pollutants and effects on health and welfare" described by the federal criteria are loaded with with aesthetic, physical, and mental comfort factors that are difficult if not impossible to objectify. As a result, the extent to which criteria reflect them is a matter of judgment to be decided on the basis of principles or values--on a subjective rather than a totally scientific, or technical, or objective basis. But decided on the basis of whose values, and through the channels of what value-gathering processes? With regard to these questions, consider the statement of two Washington, D.C., lawyers:

> Before HEW . . . issues criteria, it must obtain the requisite advice. It was intended that HEW obtain the best advice available. Industry members should be prepared to do all they can to help develop sound criteria and

accurate control information. Individuals invited by HEW to be a consultant or to work with an advisory committee in either of these areas will not be acting as industry representatives as much as experts in their fields. Therefore, other members of industry not serving on such committees must also seek an opportunity to present their points of view and information. This may be done by contacting members of an advisory committee after they have been announced or in consultation with state and local officials who will confer with the federal government on the local problem [Martin and Symington, 1968: 244-45].[6]

The statement underscores the distinctly political or "value-loaded" nature of the criteria-decision process. Indeed, that process may be as much a political one as the later steps bearing on the setting of air-quality standards, which were explicitly politicized by a statutory requirement of that great instrument of participatory democracy, the public hearing.[7]

This last observation suggests a number of questions: What bearing might be conclusions reached in the criteria-formulation process have on the decision processes of later stages (whether undertaken by legislative or administrative bodies), such as those relating to standard setting and the design and implementation of enforcement programs? What might the answers to this question suggest about the degree to which the criteria-formulation process should or should not be politicized (or open to competitive values and interests)? If it should not be politicized, what adjustments are necessary? If it should remain politicized, to what extent does the present structure provide unequal access to various interests, and how do we go about constructing fairer procedures? Do the answers to these questions suggest alterations that should be made in later stages of the decision process and, if so, what are those changes, where should they be made, and how? How can the invasion of nonpolitical stages by political forces and considerations be prevented? How can the dominance of political stages by intense and often overzealous interest groups, whether they speak for industry or "the public," be avoided. (Surely it would not be unfair to conclude that it was a small and intense interest group that captured a hearing in Pennsylvania, alluded to in note F, and--just as surely, it would seem--there is no reason to suppose that the fact that that group was composed of "just plain citizens" tends to assure that the result of the hearing was more in the actual public interest than if the hearing had been dominated and captured by industry lobbyists.)

So long as the plain and simple question of how much pollution will be allowed remains one of values, answerable only by balancing

subjective considerations, questions like these are crucial ones. And
they are also questions to which the lawyer, with his insights into the
reach and limitations of process and procedure, and his knowledge of
the ways of negotiation and compromise, can contribute much, if not
by himself, then along with others from such disciplines as economics,
sociology, and political science.[8]

Unfortunately, lawyers and law-trained people have devoted very
little attention to the issues springing from the distinctly subjective
nature of environmental-quality questions. As a result, there have
been few insights into the processes by which those questions should
be decided or of the role that legal institutions should play in problem
definition. Some observers apparently believe there is agreement
about quality goals and that only the mechanics (lawsuits, legislation,
ans so forth) of how to achieve them need be decided.[9] Others recognize the judgmental dimensions of environmental problems but are
satisfied to stop there, to point out that much of our present problem
"is due largely to the lack of any consensus to effectuate the necessary
change," but to suggest only that we need a concept of living "in harmony
with nature [which] can only be created through education and persuasion" (Reitze, Pollution Control, 1969: 926-7; cf., Hardin, 1968).
But it might be fair to ask a lawyer: Education and persuasion of
whom, and by whom, through what specific means, and to what specific
ends? Such have seldom been asked, let alone answered.

It should be clear that we are not downplaying the function of
problem indentification as opposed to problem solution; one of the
great and necessary tasks of legal process is to help define the dimensions of social problems (for example, as lawsuits have done in
the case of poverty and racial injustice and as the congressional hearing
has done in the case of national malnutrition). But it is time to begin
identifying in a more specific way problems that are not so obvious.
Yes, the environment is polluted, but how clean do we want it, and
how do we go about deciding that? Yes, the public has been indifferent,
but why, and how do we turn indifference into constructive concern?
And, more particularly, at which level of government should decisions
about environmental quality be made? Which branch of government
(for example, the administrative or the legislative), should bear the
responsibility for the ultimate operative decisions about quality? At
what stage and where should hearings be held, and what sorts of
questions should they be concerned about? Does local control result
in more citizen participation, but parochial decisions? What sorts of
practical tradeoffs can be identified and evaluated in confronting these
questions? We are not suggesting that time and resources are available to explore all of these and similar issues, or that the answers
would in each instance be worthwhile. However, some inquiries like
these must be made by law-trained people. Only when questions become

AIR POLLUTION AND LEGAL INSTITUTIONS

more specific will answers become less diffuse and more relevant.[10]

AIR POLLUTION AND THE COURTS

A striking feature of legal scholarship, reflected in the body of knowledge about pollution and the courts, is its preoccupation with judicial process. We have much more insight into the role of the courts, actual and potential, in pollution problems than we do, for example, into the legislature's role. There is an irony here, for one of our clearest and most significant insights is that the courts can play only a relatively small, although important, part in the environmental drama.

The discussion that follows focuses primarily on the courts in the context of private lawsuits seeking damages or injunctive relief against a polluter, and citizen suits challenging the pollution planning and enforcement activities of public officials. Two other media of judicial intervention--the adjudication of criminal and civil charges against alleged violators of pollution-control statutes and the policing of legislative programs under the constitutional requirements of due process and equal protection--are so much a part of the process of legislation and administration that they will be considered (albeit briefly) in that context in the next section. The main concern of this section is with the courts acting out the common-law tradition of rule origination, development, and application.

Objectives of Environmental Litigation

By assigning specific objectives to litigation arising out of environmental problems such as air pollution, the strengths and weaknesses of the courts as a medium of pollution control can be assessed more systematically, and areas that need change or at least further investigation can be isolated. Objectives in this context are not those of individual litigants, whose primary aim is usually to win either a judgment or some sort of moral victory, but rather the sorts of objectives that litigation as an institution should serve. For example, an individual litigant's objective may be to exert pressure on an adversary by bringing a spurious but expensive lawsuit; but this is most surely not an institutional objective of litigation. Accordingly, barriers are erected to guard against such suits, and various penalties may be imposed if they are discovered. The discussion that follows assigns three institutional objectives to pollution litigation; while the list is not all inclusive, it represents the primary aims of importance here.

Internalizations of Costs

A common analysis of the air pollution problem is that it represents a case where certain costs are imposed on society at large by the activities of polluters and, because of failures in the market mechanism, those costs remain external to their source--that is, they are not taken into account by polluters in deciding whether or not their activities are worthwhile to them or society at large. Since the late 1960's, some legal scholars have begun to make a constructive examination of private litigation as a means of internalizing these costs (for example, through the award of damages against polluters), thus bringing them to bear on private-production decisions (Katz, 1969; Baxter, 1968; Atwood, 1969; Wright, 1969).

Not all of the work thus far has considered the relevance of Coase's article (1960); rather, it has proceeded on the basis of the traditional Pigovian analysis on which Coase cast doubt and has, accordingly, assumed that polluters should be required to pay receptors for pollution damage. Coase's analysis, on the other hand, would ask which group, polluters or receptors, is the cheapest cost avoider. That is, in a costless market, would receptors end up adjusting in some way to the presence of pollution, or would polluters end up taking abatement steps? Lacking that information, Coasian analysis would suggest asking whether transaction and information costs would be lower with the initial right in receptors (i.e., with liability on polluters), or in polluters (i.e., with no liability on polluters). This analysis would envision "bargains" (perhaps instigated by lawsuits), or an exchange of rights between polluters and receptors, and would argue for that arrangement of rights that would minimize bargaining (transaction and information) costs, thus permitting the market to most cheaply correct any errors resulting from the initial assignment of rights. The analysis would also suggest consideration of other means, such as programs of government regulation, that might permit errors to be corrected even more cheaply than under a regime of liability rules (Calabresi, "Transaction Costs," and "Does the Fault System," 1968).

While it may be that some economists have grasped the full significance of Coase's thesis for liability rules, most lawyers surely have not. (But see Calabresi, all sources; Wright, 1969; and Atwood, 1969.) Indeed, it is not at all clear that law-trained people are even confident that they can formulate relevant questions. In some limited senses, this may not be a significant shortcoming. The few commentators who appear to have considered the question have concluded that a Coasian analysis suggests that bargaining costs are lowest when liability rests on polluters (Calabresi, "Transaction Costs," and "Does the Fault System," 1968; Wright, 1969).[11] The reasoning

underlying this conclusion can be illustrated as follows, according to
Wright (403-04): Assume that there is a single polluter, P, and a
class, C, of pollution-receptor citizens. If no liability were placed
on P (in other words, if P were given the right to pullute), the members
of C would face the expenses of getting together, holding meetings,
negotiating among themselves, and deciding what sum it was worth to
all of them (and in what proportion), to offer P in an effort to induce
P to stop polluting. These would be expensive transaction costs, and
they would be increased by the fact that there is generally a large
group of receptors (for example, all of the citizens living downwind
of a polluting factory).

Each member of the group would be likely to feel that he need
not contribute anything, as any offer to P would benefit him (as a
member of the downwind group), whether he paid or not. He could be
a free rider, for no member could be excluded from the benefits of
the group's offer to P. Because of these high transaction costs, voluntary bargains between P and C would be unlikely to occur if P had
the enforceable right. Moreover, forced bargains in the frame of
a lawsuit would also be unlikely to occur. P could enforce his right
simply by polluting. C would have no basis for a suit (P would be
within his rights), and, as explained above, no other bargain would
be likely to occur; even if such a bargain did occur, it would be only
after heavy bargaining expenditures. In short, with the right vested
in P, internalization would probably never come about.

Many of these difficulties disappear or are minimized, however,
if P is liable for polluting. Members of C would have an incentive
to sue because they could recover damages--there would be far less
free riding. Members of C would not have to meet and negotiate
among themselves and with P, for each could sue individually. Moreover, some form of bargaining would be sure to occur. P would know
that he faced a forced bargain in court and would thus be motivated
to take steps to avoid pollution damage or to make voluntary offers
to members of C, in order to avoid the costs of a lawsuit.

This analysis would appear to hold even if there were more
than one polluter, as long as there were significantly fewer pollution
producers than pollution receptors. When we assume an absence of
information about who is the cheapest cost-avoider, it seems almost
axiomatic, in the case of air pollution, that the polluter is the proper
target for liability rules, essentially because of the typical high ratio
of receptors to polluters, the polluter's greater command of relevant
information and technology, and the fact that the problem centers
around a common property resource, air, from the enjoyment of which
free riders cannot be excluded without prohibitive policing costs
(Krier, 1970: 117n. 38a). Because most persons would probably

consider liability on polluters as the most desirable outcome from the standpoint of income distribution (not to mention moral judgment), the situation presents a set of goals that are quite compatible and reinforcing.

To the extent that the fact assumptions about pollution problems which lead to this axiom hold, it appears to matter little whether liability is assigned to polluters on the basis of Coasian analysis, on the basis of traditional Pigovian analysis, or on the basis of some moral judgment about the "badness" of pollution-causing activity. But what of the case where at least some of the assumptions do apply? What of the case, for example, in which the receptor is a unique firm that requires absolutely dust-free conditions in order to engage in its specialty of assembling precision equipment? Here there are millions of polluters (for almost all of us and all of industry engage in activities that cause dust, such as cutting lawns or doing the wash), but only one suffering receptor. To eliminate the pollution at its source might be relatively inefficient, inconvenient, and generally undersirable in its impact, but the application of a rule against "polluters" under these circumstances would probably compel these results. There may be many similar kinds of situations of which we are unaware. Because liability has been traditionally assigned on bases--moral or otherwise--that have paid no heed to the way one situation differs from another in terms of transaction costs, information costs, and so forth, those differences do not tend to stand out. In short, we are not sensitized to them, yet they may be very relevant to our policy decisions.[12]

This seems to suggest that it would be worthwhile to inquire about the variety of pollution problems--problems where there are many polluters and many receptors, as with the case of the automobile; problems where there are many polluters and few receptors, as with the case of the assembly firm above; problems where there are few polluters and few receptors, as with the case of the factory in remote farmland--and about the variety of cost-avoiding techniques of receptors and polluters as to each strain of the problem. It might also be worthwhile to ask how the findings bear on the decision about imposition of liability rules. Moreover, this is an inquiry that should be undertaken not simply by lawyers, and not simply by economists, but by lawyers and economists working in concert. And, as should become clear, this can be said of much of the research that needs to be done in the case of pollution problems and legal process.

Once it is decided, however, that liability should be imposed on the polluter in a specific type of case, there is a whole store of information about how to achieve that goal. One result of the long preoccupation with judicial process has been an industrious gathering

of kernels of knowledge about how to make polluters pay. The basic
causes of action for relief by damages or injunction against polluting
activity--negligence, nuisance, trespass, and strict liability--have
received considerable attention. (See Seamans, 1970; Katz, 1969;
Baxter, 1968; Porter, 1968; Aborn and Axelrod, 1968: 217-20; Juergensmeyer and Morse, 1968: 299-306; Bellis, Kolsby, and Wolf, 1968;
Juergensmeyer, 1967; Berger, 1967; Rheingold, 1966; Kennedy and
Porter, 1955; and Steinberg; 1954.) And, the developing theories,
especially strict-products liability, are coming to be both tried out
in lawsuits and discussed in literature (Katz, 1969: 623-41; Esposito,
"Air and Water Pollution . . ., 1970: 38-40; Juergensmeyer, 1967,
n. 7; McCarthy, 1970). One theory of recovery, the inverse-condem-
nation action, has been almost entirely ignored by the lawyer and legal
scholar alike, yet it deserves the attention of each (Juergensmeyer,
1967: n. 7). Theories of strict liability call for the greatest proportion
of scholarly energy, both because they are in stages of rapid development and because they best serve the end of cost internalization by
abolishing, or at least largely avoiding, the necessity of proving fault,
a concept that ordinarily ignores the relevant considerations. Whether
or not a polluter is at fault, for example, reveals nothing about whether
he has the lowest transaction costs and seldom indicates anything
about whether he is the cheapest cost avoider.

Improvement of Private and Public Decisionmaking

Private litigation should function to encourage better private
and public environmental-quality decisions. To some extent, cost
internalization serves this end, especially in regard to private decisions: If rules of liability are fairly clear and an enterprise can
anticipate that it will be liable for damages done to the environment
by its activities, the enterprise will have some incentive to evaluate
alternatives that might reduce or avoid environmental damage, such
as choosing a different location for its activities or developing new
technology. As we shall see, this theoretical model often breaks
down in practice, and even when working smoothly, the cost-internalization function does not always serve to improve private decisions
simply because, under present law, the liability question will often
be decided without much regard to the quality of the defendant's
decisionmaking processes.[13] Take the favorite liability theory in
air pollution litigation, that of nuisance, as an example. Many courts
will decide the issue of nuisance without considering the underlying
management decisions that led to pollution production: Has the
company done research and development in air pollution control
technology? Has it considered alternative means of production, or
a different location? As a result, many companies probably ignore
the expensive process of thinking about how their activities bear on
air quality simply because, unless the thought process itself leads

to a new and cheaper means of avoiding damage liability, there is no payoff in worrying about the environment.

A larger problem is the company that escapes liability because its emissions are held not to constitute a nuisance, although the company could, through the application of knowledge within its command (but not, under present law, within the command of the court), further reduce emissions at a cost lower than the resulting gains in air quality. Largely because of considerations such as these, Katz (1969: 615), suggested that "the balance of considerations in nuisance cases may yet be broadened to include an appraisal of the defendant's search for possible new scientific discoveries or engineering designs." We see this not as a possible doctrinal development that would allow an escape from liability under nuisance law as presently applied, but as a rationale for extending nuisance to situations which it might not reach at present. To what extent could such a doctrine effectively be applied in a way consistent with protection of trade secrets? Would application of the doctrine be wise in making the trial judge (or the jury) an arbiter of the "best" or "sufficient" research and development? Would application of the doctrine be inconsistent with the objective of cost internalization, as it might tend to free the cheapest cost avoider or the cheapest bargainer simply because he did all the research one could expect? Or would losses in incentives here be balanced by the fact that while damages might not be present to work as a stimulant, the research requirements of the doctrine would? The point is, then, that while the cost-internalization function of private litigation goes part of the way toward improving private decisionmaking processes about the impact of activities on air quality, it does not perform the additional, useful function of ventilating the decision process, of exposing it to public view, and of bringing both more pressure and more information to bear on decisionmaking.

The dicussion thus far has focused on private decisionmakers. The decisions and activities of public agencies also bear heavily on environmental quality, most notably in regard to the enforcement of control regulations and the planning of resource allocations (e.g., planning the size, number, and location of power plants in the area under the planning agency's jurisdiction). Here, too, privately initiated litigation can contribute much to the improvement of public decisions as they bear on air quality. (Jaffe[1961 and 1968] , termed this sort of privately initiated suit a "public action," in light of the suit's focus on the vindication of the public interest rather than the individual rights of particular plaintiffs.) Investigation and development of plausible legal theories is especially needed with regard to enforcement suits. For example, citizens dissatisfied with laxity in the enforcement activities of air pollution control officers have little effective judicial recourse under present law. They might seek recovery against the allegedly nonfeasant official for negligent failure to abate a public nuisance, or they might ask for a writ of mandamus compelling the enforcement of abatement statutes, but with the law as it now stands, either strategy would be likely to fail. It has been argued, however, that the law in each of these areas is ripe for change (Esposito, "Air and Water Pollution . . .," 1970: 41-45).[14]

As an alternative strategy, the private citizen might opt to bypass the public-enforcement official by attempting to clothe himself with that official's unexercised authority to abate public nuisances. Here, too, the plaintiff is likely to fail, for he must not only allege but prove that he has suffered "special damage" above and beyond that suffered by the public at large, and this is often difficult. In several states, however, statutes specifically provide for special proceedings by private citizens to abate nuisances in the name of the state (Katz, 1969: 622).[15] However, Jaffe suggested that the concepts now developing in standing decisions (discussed below), might play some part in a revision of the law of mandamus and public nuisance (Jaffe, 1970).

Citizens dissatisfied with public agencies' planning activities might seek participation in proceedings incident to the planning function (such as administrative hearings), or seek judicial review of agency planning decisions that allegedly failed to comply with statutory guidelines. An important body of law is rapidly developing to the effect that private citizens have standing to participate in and seek judicial review of many such planning decisions if their past activities and conduct demonstrate that they have a special interest in the subject and outcome of the proceedings.[16] This development is probably of more importance with regard to federal than to state planning agencies, for the state courts have long granted standing to taxpayers to complain about public actions allegedly illegal and involving the expenditure of public funds (Jaffe, 1970).[17] As Jaffe pointed out, however, the impact of these federal decisions might be most felt at the state level in the liberalization of public nuisance and mandamus law (Jaffe, 1970).[18]

Defining New Issues

Environmentalists in the trial bar have become quite sensitive to the fact that litigation might serve the objective of spotlighting areas which need further thought and attention from the public and lawmakers. One finds this attitude reflected in the following:

> Court actions now draw the most attention. Citizen groups go to the courts as a last resort when legislation, governmental executive decisions, and public opinion have failed to halt actions which they believe are endangering the environment.

One of the nation's leading trial lawyers in the environmental field, Victor J. Yannacone, Jr., has stated: "Every piece of enlightened social legislation that has come down in the past 50 or 60 years has been preceded by a history of litigation in which lawyers around the country have focused forcibly the attention of the legislature on the inadequacies of existing legislation."[19]

Litigation can uncover administrative abuses, gather together hard evidence about previously unknown or poorly understood problems, pinpoint shortcomings in present laws, and suggest the shape and direction of new laws. Often these functions are merely tolerated as incidental to settlement of the particular controversy before the court; that they may also have an impact on larger questions is not (at least formally) considered as a justification for their operation. Accordingly, the rules of evidence will usually confine within narrow boundaries the amount and kinds of information that may be introduced. In some circumstances, however, the function of providing information bearing on interests and ends larger than those immediately before the court is accepted as legitimate.[20]

But a judicial hearing is not a legislative hearing; under the most permissive of circumstances it is a more restricted system, both less searching and open to fewer points of view. As a result, there are dangers of distortion if litigation-generated facts are translated too literally into information that shapes broad public-policy decisions. Consider the following taken from an account of an environmental conference:

We have been using the lawsuit as a device to teach people that they can demand environmental controls and get them, [Professor Sax said].

Professor Sax's educational tactic did not meet with complete favor at the table, particularly among the economists. Professor Breton remarked that he thought it wise to differentiate between the situations "where in one case the pollution is serious enough to arouse the public, and where in another we have to 'educate public opinion.'" He then warned: "We should be careful not to engage in calling something education which, when other people try to sell us soap, we call advertising" [Shapiro, 1970].

Shortcomings and Palliatives

The discussion thus far has made at least implicit some of the shortcomings of litigation, of the courts as an institution for effective pollution control, and some possibly desirable directions for investigation and change. What follows is a more systematic critique of those shortcomings and of the sorts of palliatives that exist or might be produced to contribute to a more effective institution. For purposes of discussion, two classes of problems can be identified: (a) problems of proof, and (b) problems inherent in the judicial process.

Problems of Proof

The problems of proof confronted by the plaintiff in environmental litigation have been catalogued, and it is enough to capsulize them here (Aborn and Axelrod, 1968; Baxter, 1968; Hines, 1966; Committee on Pollution, 1966: 231-33).[21] The problems arise from the plaintiff's burden to establish intentional or unreasonable conduct as a basis for liability under many of the theories common to air pollution litigation and, even when that burden does not exist (as in the case of strict-liability theories), to prove the casual link between the allegedly wrongful conduct and the injury of which plaintiff complains. There are, of course, well-known doctrines that lighten these burdens in some respects, but they by no means do a systematic job. Res ipsa loquiter, for example, permits an inference of negligence upon a showing that the event causing damage is of a kind that ordinarily does not occur in the absence of negligence, that the event was caused by an instrumentality within the exclusive control of the defendant, and that the damage was not due to the plaintiff's voluntary action. The doctrine has found application in pollution lawsuits but hardly on an across-the-board basis. [22]

Similarly, the doctrine of negligence per se can be a valuable aid to the plaintiff in proving liability. According to this doctrine, the provisions of a statute, ordinance, or administrative regulation may be taken to establish the standard of required conduct if the purpose of the provisions is, exclusively or in part, to protect a class of persons that includes the plaintiff, to protect the particular interest of the plaintiff which was invaded by the defendant's conduct, and to protect that interest against the harm which resulted and against the hazard which produced that harm.[23] Most courts hold that when the above conditions are met, an unexcused violation of the provisions is conclusive on the question of negligence, but in some jurisdictions the violation is treated only as evidence of negligence. Which of these views is applied might turn on whether a statute, an ordinance, or an administrative regulation is involved, for there are courts which follow the majority rule in the case of violations of the former but regard violations of the latter as evidence only (Prosser, 1964: 202).

The value to the plaintiff of the doctrine of negligence per se will obviously grow as more and more detailed, air pollution control legislation is adopted (see Borchers and Miller, 1970). But neither this doctrine, nor res ipsa loquitur, nor theories of strict liability remove all the obstacles to successful litigation. Cause and effect must still be established, and the problems of proof here can be intractable. It will often be the case, for example, that multiple-emission sources contribute to a locality's air pollution problem, or that the type of injury involved in the lawsuit (e.g., damage to the plaintiff's health), could have been produced by conditions other than foul air. Even for problems such as these there are palliatives, but they are at best a mild tonic.[24]

The tendency of the courts has been to reallocate the burden of proof regarding liability in roundabout ways, rather than directly. The result has been doctrines--such as res ipsa loquitur, negiglence per se, and so forth--that have been less than systematically applied. The air pollution case in which res ipsa loquitur applies, and the one in which it does not, for example, often will not differ in any way so fundamental as to call for different judicial approaches to problems of conserving the air resource. Indeed, conservation litigation has a fundamental constant that the legal system has been slow to recognize; once its implications are seen and understood, it should be possible to consider a more broadbased and systematic reallocation of the burden of proof.

The constant in resource-conservation litigation is this: The person or enterprise whose activities deteriorate a resource will always be the defendant. This is because, in the absence of legal constraints, conducting a resource-deteriorating activity can foreclose the conduct of nondeteriorating activities, while the converse is never true. The polluter's conduct can stop people from breathing clean air, but the breather's conduct cannot stop the polluter from going about his dirty business. To obtain relief, the breather must go to court, and it is one of the plain facts of the judicial system that plaintiffs carry the burden of proof on most of the important issues in a lawsuit, not to mention the financial burden of initiating litigation. Thus, even when substantive rules express a policy in favor of resource conservation, basic procedural mechanics tend to bias the system in the opposite direction. It is important to begin considering ways to make systematic reallocations of the burden of proof in resource-conservation litigation to counteract the present discrimination against conservation (Krier, 1970).

Problems Inherent in the Judicial Process

Some of the reasons why judicial process can never be a completely satisfactory institution for pollution control are inherent in the process itself. We shall sketch a few of the most fundamental of these and briefly discuss some existing and possible palliatives.[25]

Lawsuits are expensive and ordinarily culminate when someone is seriously injured in pocket, person, or principle. However, pollution damage is often diffuse; a great number of persons may be injured, but each to only a small extent. Although social costs may be high, individual damage will ordinarily be so low that an individual is discouraged from hiring a lawyer to prosecute a suit. As a result, costs are internalized to only the most modest degree; litigation thus fails to realize one of the primary objectives we have assigned to it.

One possible (and probably impracticable) way to guard against this failure is to have the judge attempt to total all the damages caused by an activity. In cases where it is perfectly clear that if all damages were internalized, the enterprise engaging in the activity could not pay them and survive, the judge might grant an injunction. Too often, however, judges probably grant injunctive relief when relief by way of damages would be more efficient (Calabresi, 1961: 535). At times, too, they grant injunctive relief in a theoretically undesirable form. For example, in Renken v. Harvey Aluminum, Inc., 226 F. Supp. 169 (D. Ore. 1963), the court ordered Harvey to install the same type of pollution-control equipment that had been used by Reynolds Aluminum, apparently on the theory that what was good for Reynolds was good for the industry. If Harvey had simply been ordered to reduce its emissions to the desired maximum, it might have found equipment which did that job at less cost than the Reynolds equipment.

The class action is an obvious although limited solution (Wright, 1969). Rule 23 of the <u>Federal Rules of Civil Procedure</u> provides a liberal class-action rule, but its value as an internalization device has been severely limited by a ruling of the U.S. Supreme Court (Snyder v. Harris, 394 U.S. 332, 1969) that class members may aggregate individual claims to reach the jurisdictional minimum of the federal courts (over $10,000) only when they share a common and undivided interest--an unlikely occurrence in air pollution litigation. While litigants may still resort to state class-action rules, some commentators find these "even less viable than was the federal provision" before the Supreme Court's decision (Esposito, "Air and Water Pollution . . . ," 1970: 36). Moreover, it may well be that the class form of action itself is not a sufficient spur to effective litigation.[26] Much of the promising experience with class actions under the federal rules has been built upon the antitrust laws which provide for treble damages to those injured. Perhaps the analogue to triple recovery in pollution suits will be punitive damages. At least one court expanded that concept to apply to a polluter who conducted activities knowing that there was a probability, or at least a danger, of harm yet failed to take appropriate measures to minimize damage.[27] The potential reach of punitive damages as a therapeutic device subject to judicial development should be explored. The existence of federal records can also serve as an independent incentive to bring class actions, as has been the case with regard to such actions by small investors in the securities markets (Wright, 1969: 418). Attorneys can make valuable use of federal air pollution documents in a variety of ways.[28]

But the most fundamental restraint on the future expansion of the class action in pollution litigation might be that class actions at the breadth of their potential would bring to the court more than the judge could handle. In 1969, a class action in Los Angeles was filed on behalf of essentially all the people in Los Angeles County against 291 corporations allegedly responsible for air pollution in the area. The court dismissed the action for several reasons, one of them being the following:

> [A] court of equity lacks facilities or competency to undertake the problem of abating air pollution within the Los Angeles Basin. [Expanding on this point, the court stated:]
>
> It is readily apparent that the control of the emission of air pollutants is a highly complex problem. . . . The court does not have the facilities to undertake the balancing of the interests of the inhabitants of the Los Angeles Basin against the needs of productive industry in this same area.[29]

It is difficult to fault the judge for this view (certainly his candor is admirable). As we have seen, the question of air quality is an essentially political one, ideally to be decided by democratic, political progresses. The judicial process is not one of these. Moreover, the courts lack the purse power and jurisdictional base needed to ferret out the facts and balance the consideration essential to a reasoned decision and to police that decision once made.[30] Surely the courts can adjudicate some class actions, but not those that seek solutions to the most serious problem, i.e., multiple-emitter pollution in large, urban areas. As to this problem, the courts will serve best to the extent that they vigorously insist on reasoned and neutral processes of decision and enforcement by other agencies of government. To even hope for much more on the part of the courts would require the most fundamental changes in judicial process--changes that would take much in time and trouble and that might result in an institution far less desirable from other than the perspective of pollution problems. It is often easier to turn to other institutions than to change deeply rooted traits in the character of the courts, and this seems true in the case of pollution. Surely the courts have a role to play, but it is not an exclusive and probably not a major one. The bulk of the burden must be carried by other agencies of government.

POLLUTION CONTROL LEGISLATION
AND ADMINISTRATION

One way to begin a survey of air pollution control legislation and administration in the United States is simply to note that government at all levels has had an almost neurotic fixation on regulation. As discussed below, government can attack pollution in essentially three ways: (a) by regulation, i.e., proscriptions and prescriptions backed up by punitive measures; (b) by subsidies or awards; and (c) by charges, or prices, or effluent fees (Mills, 1966: 43-44; and Dales, 1968). There has been an incredible amount of regulation, a touch of subsidization, and virtually no pricing.[31] And it is fair to say that most of the significant gaps in knowledge about what government can and should do with regard to pollution problems are the direct result of this regulation imbalance. There is little empirical information about the merits and demerits of subsidies and awards and virtually none about pricing. Because there has been little in the way of a comparative institutional approach to solving pollution problems, systematic studies about the advantages and disadvantages of various control patterns are scarce.[32] Our approach has been monolithic--and barren.

What accounts for this "great American regulatory tradition?" A few answers seem at least plausible. It has been suggested that the history of air pollution control in the United States illustrates the "catastrophe theory of planning," the direct and immediate response by government to dramatic incidents (such as the disastrous Donora, Pennsylvania, episode, in which 20 persons died and almost 6,000 became ill as a result of a three-day period of excessive air pollution).[33] Such a theory may expose the roots of the regulatory tradition. Governmental response to crisis is often typified by quick and, therefore, generally crude measures aimed at preconceived "wrongdoers" like industrial polluters.

Consider the following comment by a Vice President of the Boston Edison Company, reported in Goldner (1966: 150):

> What people fail to understand is that a utility merely represents one large power-generating operation in lieu of each home having its own power-generating and heating devices. The small quantities of smoke which would be given off by each of the multitudes of homes if they used their own power-generating devices are now consolidated into emissions from a few plants and so become a noticeable target for attack.

Governmental response is designed to punish the wrongdoers and placate intense public concern. The goal is immediate relief and clear results. The prohibitions and penalties of regulatory programs best suit all of these ends. With time and experience, it may become clear that an existing program, conceived out of crisis, is unsatisfactory, and that more subtle measures are necessary. But the path of least resistance--and, often, least short-run cost, a matter of importance to politicians--is to improve the old, rather than abandon it for an entirely new approach to government intervention. It may be, too, that the roots of the regulatory tradition lie in the fact that most people, including most public decisionmakers, probably do not understand either the ways in which social and economic systems function to produce pollution problems or the programs of control offered by theoreticians and sophisticated experts. Then too, people, including many theoreticians and experts, may simply have concluded that regulation is the best approach (see Hagevik, 1968: 376-78).

Whatever the reasons, the regulatory tradition is ours; it has been with us for years and will probably continue well into the future (Hagevik, 1968: 376-78). It has provided some valuable lessons. The most beneficial application of the fruits of those lessons, however, may call for quite new responses by government. And the regulatory tradition is the constant which the force for change must overcome.[34] To the extent that this rigid tradition is the product of crisis, it suggests that at least some of the energy of crisis should be harnessed and directed toward the establishment of low-visibility study bodies whose job it would be to conduct research and prepare alternative programs of government control. Those programs should be introduced into the legislatures only when the panic of the moment has passed. Perhaps, in short, it would be best to adopt a tactic of striking not when the iron is hot, but when it is cool.

It is beyond the purpose of this chapter to discuss all the varieties found in the jungle of legislative-control programs. We propose rather to outline the main themes of regulation, subsidies, and pricing in a broad and comparative fashion in order to suggest what is known and to consider the research implication of what is not.

Regulation

Regulatory programs have appeared in a variety of styles and, when carefully drafted, have almost uniformly withstood constitutional attack. (See Delogu, 1969; Edelman, 1965; Kennedy, 1957, 1961, 1962, 1963, 1964, 1966; Kennedy and Porter, 1955; Pollack, 1968; and Goodman, 1956.) This does not mean, of course, that constitutional and other legal constraints have no impact on legislative design and the style and success of enforcement. As the following discussion illustrates, quite the contrary is true.

Regulatory techniques for controlling such problems as air pollution can be placed in two broad categories--performance standards and specification standards.[35] In essence, performance standards allow polluters to act as they wish, so long as their emissions do not exceed specified limits for particular pollutants. Early legislation of this sort was aimed at the density and opacity of smoke (measured against a visual chart, the almost proverbial Ringelmann Scale), but modern statutes commonly contain further limitations on particular pollutants, such as particulate matter, sulfur dioxide, and nitrogen oxides. Specification standards decree not what goal must be accomplished but how it must be accomplished. Most commonly, they require pollution-control equipment that meets certain design requirements, but they may go further and require particular makes and models of equipment. (Note [e], 1968.) Laws forbidding certain activities, such as open burning or the use of certain fuels, can be considered types of specification standards, because they go beyond simply establishing an emission goal.

Regulatory programs can, and at times do, employ both performance and specification standards. They might, for example, set an emission (performance) standard for smoke, prohibit fuels that produce inordinate amounts of smoke, and establish specification standards to control other types of pollutants. Some control programs blend the two approaches in another manner. They establish a performance standard but require that operations cannot begin until the control agency is satisfied that the source is designed in such a way that it can be expected to meet the specified emission standards. Programs of vehicle-emission control have employed both specification and performance approaches. Some, that is, insist on the installation of certified devices capable of achieving given standards; others (most notably the federal program), simply instruct automobile manufacturers that the given standards must be met, then provide a means for assembly line or on-the-road testing.[36]

Performance standards have one distinct theoretical advantage over specification standards: by leaving to the owner of a pollution source the decision of how best to comply with the law, they tend to induce him to do so in the most efficient manner (that which is cheapest for him), because it is in his interest to develop and employ least-cost solutions (Walker, 1968: 85). This observation, however, applies in only a very limited way to the case of control aimed at many small polluters, such as people using backyard incinerators or driving automobiles. Small polluters have little capability to discover and design the devices best for themselves. They can, of course, save by purchasing on the market the cheapest effective control equipment, but the administrative costs of policing a large number of small operators to see if their emissions meet specified limits would probably be higher than the total of individual savings. In such cases, if a regulatory approach were to be used at all, then some sort of specification program would seem justified.

This last point makes it clear that the efficiency of a control program cannot be judged from the perspective of the regulated class alone. Relative costs of administration and enforcement must also be considered. While the question is a complex one, performance programs are probably more efficient from this perspective as well (with the exception of the small polluter). Specification programs are ordinarily administered through permit systems that require submission of specification data to an authority before operations begin (Walker, 1968: 84). Activity cannot lawfully begin without a permit, and no permit will be issued until the authority is satisfied that the required standards are met. The permit system itself requires a large staff that must engage in time-consuming appraisals and negotiation sometimes ending in the granting of a variance, which is often nothing more than an arbitrary escape hatch for the most persistent or powerful of the regulated class.[37] Moreover, there is some opinion that staff work under a permit system tends to be overcautious, resulting in unnecessarily expensive overengineering of control equipment (Walker, 1968: 85). The analogue of a permit for equipment already in operation is a certificate. Sources of pollution are periodically tested during operation to see that they meet design and operational standards; if they do not, a certificate of operation will be withheld. Administration of a certification program requires an even larger staff than the permit system and has the additional inefficiency of lost time and labor brought about by interruptions in business activity while machinery is being tested and evaluated (Note [e], 1968: 242).

Of course, against these inefficiencies of specification programs, those created by a performance program's need for policing emissions at the source must be weighed. In most cases, however, only spot checking is necessary, and it is unlikely that the costs of this procedure are significantly higher than annual-certification costs. Even specification systems call for policing measures; control equipment after all, can, break down or be ignored. Reasonable conjecture and some evidence suggest that performance programs require no more inspectors than specification programs and relatively fewer engineers and other personnel.[38]

The relative inefficienceis apparently inherent in a program of specification standards are compounded by the fact that, ordinarily,

AIR POLLUTION AND LEGAL INSTITUTIONS

such standards are enforced by criminal process, probably the most cumbersome coercive tool available.[39] The alleged violater receives all the constitutional protections that apply to any criminal trial. A study in Boston (Kovel, 1968: 154), concluded that the criminal-enforcement scheme "has not been of much help for since its inception, there has been an average per year of 700 to 800 complaints, 4 prosecutions, only 1 conviction, and no noticeable improvement in air quality." (but see Mix, 1968.)

In terms, then, of both administrative costs and compliance costs, performance programs appear to be relatively more efficient than specification programs in reaching a collectively establish air-quality goal. Neither system appears to leave much room for market correction, i.e., neither appears to leave much room for bargaining in any ordinary sense.[40] To some extent, the existence of a mixed system of regulation and liability rules might tend to correct for this rigidity. As long as the air-quality level required or achieved by a regulatory program is lower than that which would be established in a costless market, the present system of liability rules might function, in a small way, to raise it.[41] But it is difficult to conceive of similar "market" corrections if the legislatively established level were too high and were rigorously enforced. Insofar as regulatory programs are concerned, all of the above remarks suggest the wisdom of opting for the relative efficiencies of performance standards, because, in general terms, they bring no apparent disadvantages. This reveals little, however, about what such a program should look like, or how it should be administered, and it is here that much work remains to be done. As Mandelker (1970: 207, 209), said:

> From the administrative side, it would seem clear that we need a prescriptive model of an effective method of administration before we can decide on what standards to choose. [Yet, lawyers] do very little work on the theoretic problems of modeling different types of legal controls, whether in a descriptive, predictive, or prescriptive sense.

This is the frontier of research for those concerned with the relations between pollution problems and the legal process. What practical, political, and legal constraints bear on program administration, and what do they indicate about the shape a program should take? Which constraints are most important in their impact, and which are most pliable? What are the net gains and losses from various approaches? Little work has been done with regard to questions such as these. It must be interdisciplinary work and, most surely, it must have the input of lawyers.

Subsidies

This and the next section deal with two fiscal alternatives to regulatory programs--subsidies and pricing. Both share with regulatory programs the feature that they are intended as means to reach a collectively determined air-quality goal. Both, as a matter of theory, are almost identical from the standpoint of efficient resource allocation; the only difference is in the initial locus of the right to use of the air resource. For example, a subsidy to polluters would recognize this right in the polluter and work as a payment to the polluter to give it up. A pricing system, on the other hand, would in essence recognize the initial right in receptors and would require payment by polluters to use the receptors' rights in the air resource. As we shall see, however, subsidy and pricing systems differ when theory is reduced to practice.

The principal subsidy programs in the case of air pollution control have taken the form of tax incentives--investment credits and accelerated depreciation.[42] The idea, of course, is that such incentives encourage efforts to achieve a higher level of air quality. However, the idea breaks down in practice. In many cases, the incentives merely reduce the cost of unprofitable equipment, and reducing the cost of such equipment does not induce a firm to install it (Mills, 1967). Arguably, however, the tax breaks might have some desirable effect by reducing a firm's costs of achieving an image as a nonpolluter (but see Roberts, 1970: 1532). Tax incentives also tend to be ineffective in that, at least in their present form, they subsidize only the capital costs of pollution control, which may be only a fraction of a firm's total air pollution control costs (Working Committee on Economic Incentives, 1967: 27). A system of award payments--direct-cash subsidies as opposed to tax relief--could avoid this problem. Payments could be geared either to the amount of abatement achieved by a firm or to the achievement of a certain standard of air quality, with graduated payments for various levels of performance. However, the measurement devices needed for a reliable program of award payments probably do not exist.

Incentive programs are also difficult to justify on grounds of economic efficiency. First, it is difficult to determine from an administrative standpoint just how much of a plant's investment should be charged to pollution control and how much to capital improvements that would have been made in any event, e.g., to expand production capacity. A program that rewarded and underwrote improvements which would be made in any event (and which, indeed, might actually work toward undesired ends) would be most efficient. Inefficiencies would also result if the incentive program encouraged capital improvements when undiscovered lower-cost approaches (such as the use of different fuel) existed (Gerhardt, 1968). Cash awards would not cure the problem. They might encourage a firm to search for the most efficient means of improvement, but the firm would still pocket the whole award, even if it were greater than the cost of improvements. Indeed, pollution abatement might itself tend to become a profitable business and could encourage the formation of firms that would engage in the wonderful business of getting paid not to pollute (Dales, 1968: 87). A related problem grows from those cases where pollution control might yield valuable byproducts that could be sold at a profit by the polluter. Do we wish to subsidize profit-making ventures? (Gerhardt, 1968.) Neither award payments nor tax incentives function to internalize the social costs of pollution on polluting firms and their consumers. The cost of both forms of subsidy are borne by taxpayers and thus discourage a shift in the market away from goods whose production involves high-pollution costs. As a result, more of such goods might be produced than would be the case if the market were allowed to function (Roberts, 1970: 1535-36).

Award payments would avoid another set of problems peculiar to tax-incentive programs. The tax system exists primarily to raise revenues in a fair way, and tax incentives erode the tax base, often in an unfair way. Moreover, once embodied in the tax laws (which are already badly pocked with scars of past subsidies), tax incentives tend to remain there. This is both worrisome, and likely, inasmuch as tax subsidies do not receive the periodic review and approval of legislative bodies given to direct expenditures and, in any event, cannot be evaluated with precision because they consist of incalculable amounts of revenue foregone (Gerhardt, 1968: 363; Note (f), 1968: 1309-11).

Most people would also find subsidies difficult to justify in terms of other goals, namely, income distribution and general grounds of equity and morality. Not only do subsidies acknowledge a right to pollute that is repugnant to most of society, but they also tend in many cases to transfer wealth to relatively affluent groups. All in all, subsidy programs seem unattractive from the standpoint of a resource-allocation goal (although they tend to be politically attractive to the extent that big business, which has carried on a long love affair with subsidies, has influence over the legislative body).

The arguments against subsidies may, however, be diluted or outweighed by other considerations in certain cases--for example, payment of subsidies to low-income persons to enable them to afford the costs of installation of pollution control devices in automobiles.

Pricing

Like subsidies, pricing programs are a fiscal means designed to achieve a desired state of air quality. Rather than positively awarding polluters for reducing emissions, they charge (or tax) the polluter for each unit of emission. Pricing is thus a direct attack on the external diseconomies of pollution resulting from breakdowns in the market system. The underlying principle of pricing is clear.

> If you charge a person for disposing of his wastes he will find ways to reduce the amount of wastes he disposes of, and . . . the more you charge him the stronger the incentive he will have to find some less damaging method of disposing of his waste [Dales, 1968: 81-82].

To put the matter very simply, an air pollution pricing system might work roughly as follows, according to Ruff (1970): An administrative agency would be given the responsibility of identifying pollutants and their sources and establishing methods of estimating emission outputs at the source. With this information in hand, the agency would designate schedules of prices for the various pollutants. Ideally, prices would vary depending on how dangerous ("costly") a particular pollutant might be, the location of the source (emissions in a desert would probably cause less total damage than in a city),[43] the season of the year, daily meteorological conditions, and so forth. The prices would be based on units of emission and would be uniform as to all polluters sharing identical circumstances.

Pollution producers would be free to adjust as they chose to the prices imposed on them: A government regulator would not be looking over management's shoulder and giving (often ill-conceived) advice. This drastic decentralization is considered one of the great advantages of a pricing system (Hagevik, 1968: 371-72). Polluters could change fuels or production methods, move to a different area, install pollution-control devices, pollute only during certain seasons, pay the price, or mix all of these alternatives in the way most advantageous to them. Consumers, of course, would tend to shift their purchases away from goods to the extent that those goods increased in price, and it is this change in demand which would encourage each firm to act in the way most efficient for it, in light of its presumed objective of profit maximization. The same phenomenon would mean higher consumption of goods, the production of which involved little pollution and less consumption (and, thus, production) of those which involved relatively great amounts of pollution. Some firms might find it cheapest simply to pollute and pay, passing the cost on to their customers or shareholders. If this were common, it would probably suggest that prices were not set high enough to begin with; they could always be adjusted

AIR POLLUTION AND LEGAL INSTITUTIONS 119

(up or down) in light of experience. A willingness on the part of customers to pay the price might also suggest that society values certain services and products so highly that it is willing to take the pollution that goes with them. This information could be valuable to policymakers.

Some economists appear to believe that a pricing system would have lower administrative costs than alternative approaches (Ruff, 1970). But others conclude that these would be about the same in any sophisticated program (Dales, 1968: 87-88, but also see 92-93). Neither argument can go very far. Since there has been very little practical experience with pricing, reliable judgments about comparative administrative costs cannot be made. Even if costs were about equal, theory would seem to suggest that pricing would be relatively more efficient for essentially two reasons. First, if we assume that firms are profit maximizers, each will act to keep its costs at a minimum. Second, emission prices can be raised until the collectively determined quality standard is reached. When we put these two reasons together, the desired quality can be achieved at the lowest cost.

A variation of the pricing model would establish a pollution ceiling but would then translate that maximum figure into a number of pollution rights (very much like shares of stock). Anyone would be permitted to purchase these rights at auction; conservation groups, for example, could buy and "bank" the rights in order to diminish pollution. Auctioning would tend to establish a price for the rights that measured the amount the public was willing to pay to diminish pollution (Dales, 1968: 93-97). An important advantage of this auction model as compared to a system of emission fees is that it requires no governmental decision about the set of prices needed to attain the desired level of quality. That decision is an exceedingly complex one, open to much room for error, and perhaps not as subject to easy correction as some proponents of emission fees seem to assume. The auction model depends only on the simpler calculation of the number of pollution rights to be issued. The market sets the prices. Moreover, the auction model would appear to be more capable of dealing with firms that do not strive to maximize profits, with firms that enjoy a monopoly position in the market, and with regulated industries--such as public power utilities--which are guaranteed an annual rate of return. Under a system of emission fees, firms and industries such as these might simply absorb or pass on the added costs of the fees without changing, in any significant way, activities that contribute to the pollution problem. Emission levels could remain high. Under the auction model, the number of rights issued would set absolute limits on the total amount of pollution. Once all the rights were purchased, pollution levels could not increase no matter how the costs of the rights were distributed. Finally, the auction model also appears to bring market forces to bear on the problem of error

correction more than does the method of emission fees, as rights could
be more easily transferred between polluting and nonpolluting interests.
The appearance might be largely illusory, however, because nonpolluting interests such as conservation groups would have difficulty
raising the necessary cash to compete for the rights. Even people
who desired less pollution would tend to take the position that they
need not contribute to the cause in the hope or belief that others would
purchase and bank the rights--an act that would benefit contributors
and noncontributors equally (Krutilla, 1967: 781-82). In any event,
market forces could only work to correct an error in setting the airquality level too low, not one in setting it too high. Proponents
of all pricing models, however, contemplate that periodic adjustments,
based on experience, would be made in the level of prices or the number
of rights. They argue that this relative flexibility in any pricing model
is one of its chief advantages, but see Roberts (1970: 1555).

There are, among economists, a number of arguments waging
over the subject of pricing, most of them growing from disagreements
about economic theory. (For summaries, see Hagevik (1968: 370-74),
Wolzin (1968: 233-37), and Wright (1969: 399-401).) Lawyers, on the
other hand, have paid very little attention to a price system as a means
of pollution control. From time to time, references are made to the
difficulty one would face in drafting a program which could pass constitutional muster, but they are unconvincing.[44] Why, for example,
would arguments urging vagueness or denial of equal protection based
on the deficiencies of measuring equipment have any more force as to
pricing programs than they have with respect to regulatory programs
which depend on, and lawfully use, the same imprecise equipment?

Various pricing models have apparently not been compared on
the basis of legal, administrative, and other policy considerations.
For example, it seems apparent that the auction model would involve
lower-policing costs than would emission-fee models. The latter call
for constant monitoring; the former would seem to require no more
than spot checks by the authorities to ensure that pollution rights
were not being abused. Is this apparent advantage of the auction
system a workable one? In another vein, how can a workable system
be constructed to bring the price system to bear on receptors as
well as polluters, so that they too have an incentive to avoid pollution
damage? What is needed with regard to pricing as a means of control
is some empirical data, some practical experience. We might find
some of that experience already in existence if we once sat down and
actually tried, after studying all the existing theoretical research, to
draft a system and a means for its administration. For example, much
of the practical experience under regimes of performance and specification standards would bear directly on the shape of a pricing program. That experience has yielded at least some knowledge about

management behavior and administrative efficiencies with regard to
air pollution control. Actually implementing a program on a pilot
basis would also generate empirical data. If that should sound far-
fetched, call to mind the fact that today's existing leased-housing
program grew directly from a pilot project and that there are currently underway the Office of Economic Opportunity's pilot programs
on the negative income tax.

Theoretical disagreements about the merits of pricing will probably go on forever, and unless some concrete steps are taken, such a
system will never be given a fair chance.[45] That would be a shame
indeed, for in the few cases in which pricing has actually been used
on a full-scale basis (for example, to control water quality in Germany's
Ruhr valley), it has met with success (Kneese, 1965). The "great
American regulatory tradition," on the other hand, has been a disappointment at best.

CONCLUSION

We have tried to assess the state of knowledge with regard to
the role of legal institutions in coping with pollution problems and to
suggest, in some general terms, areas badly in need of development
and research.[46] There are several matters on which our discussion
did not touch.

One concerns the question of international institutions to cope
with the fact that pollution respects no political boundaries. (See,
generally, Kennan, 1970; Adinolfi, 1968; and Rempe, 1968.) Another
concerns the environmental quality study and advisory councils being
created, either by executive or legislative mandate, at the federal
and state level.[47] These bodies are charged with the responsibility
for determining and formulating environmental policy, and, to some
extent, insisting on its pursuit. Among other things, the councils illustrate the function of government as a gatherer and disseminator of
information; for the most part, their activities have just begun, and
there has been little opportunity to make inquiries into their work.
One study is reported in Krier (1971). It is important that such inquiries begin as soon as possible, for the councils are likely to have
a broad impact on environmental planning and to raise significant
issues of legal process. For instance, what latent coercive powers
do these study and advisory bodies have? How democratically do they
function? How do they see and implement their missions?

We have also left virtually untouched two matters closely related
to each other. The first concerns the roles of different levels of govern-
ment--federal, state, local, and regional--in environmental-quality

control. While regional programs are the subject of much current discussion, the operational meaning of the concept and the extent of our commitment to it are by no means clear. For example, federal law provides for air pollution control regions in the case of stationary sources, but program implementation and enforcement are left primarily to the independent acts of the states within any particular region; moreover, the states in turn delegate enforcement to local governments. More attention should be devoted to the optimal arrangement of responsibilities of different levels of government with regard to environmental-quality control. (See Schueneman, 1968: 759-67; Kennedy, 1966: 131-33; Kennedy, 1959: 389 Proceedings, Third National Conference on Air Pollution, 462-524.) Specifically, there is a clear need to investigate the optimal mix of responsibility, not simply in general but for different sorts of control programs, be they direct regulation, subsidies, or pricing. Pricing systems, for example, may make regional programs politically more acceptable by reducing the number of details on which stubborn, independent sovereigns within a region must agree.[48] Any program of control should be assessed in light of alternative levels of government, and the role of any government should be assessed in light of alternative-control mechanisms.

Proposals for regional-control programs tend to go hand-in-hand with proposals for an integrated-agency approach to environmental management. (See, e.g., Roemer, Frink, and Kramer, 1971.) While a mild movement in favor of such an approach appears to be underway, there is as yet little information bearing on the design of integrated, environmental agencies (Herfindahl and Kneese, 1965: 91-92).[49] While we have put explicit consideration of this issue beyond the scope of this chapter, it does seem clear that the discussion of pollution-control legislation and its administration applies as much to the design of an integrated agency as it does to separate agencies concerned with distinct categories of environmental problems. In any event, disparate programs aimed only at individual problems will be with us for some time, and there is much that can be done while we wait for the grand approach.

RESEARCH RECOMMENDATIONS

We have tried to identify a number of areas that seem to call for the immediate attention of those concerned with pollution problems and the legal process, and they can be summarized briefly. The role of legal institutions in assuring the sort of democratic decisionmaking processes so necessary to sound decisions about environmental quality is of great importance. There must also be continuing work into property rights as the interface between law and economics and into the bearing of recent insights on judicial and legislative schemes.

For example, the full, practical relevance of Coase's theorem to the variety of pollution problems has not yet been explored. Research from the standpoint of pollution and other environmental problems is also badly needed with regard to the expanding frontiers of tort law, the adjustments necessary in existing restraints on citizen-enforcement suits, and the bias of burden-of-proof rules.

When we turn to matters that relate especially to pollution legislation and its administration, several topics stand out. The first is administration. We need far more data about the relationship between the rule of law and the form of administration, about the range and effect of legal constraints on a desired mode of administration. The second is that neglected orphan, pricing systems. Lawyers and economists working together should, after the necessary background work, draft a pricing program, design a plan for implementation and administration, and then attempt with vigor to peddle the product. Finally, we should pay far more attention to the ways in which forms of legislation and administration bear on the effectiveness of the judiciary. We have suggested that the judiciary is hampered in dealing with pollution and other environmental problems. It may well be, however, that many of the shortcomings in the courts are merely reflections of shortcomings in our programs of legislation and administration. Different forms of governmental intervention, such as pricing systems, might be not only more efficient in and of themselves but might bring with them the additional benefit of a more effective role for the judiciary. This point underscores, we believe, how badly we need open-minded, system-oriented research with regard to pollution problems and the legal process.

NOTES

1. See Royko, "The Fox" Stalks the Polluters . . . and Strikes, Los Angeles Times, September 23, 1970, p. 1 col. 1; see also "The Fox", Newsweek, October 5, 1970, p. 90. The government monopoly of force underscores the inaccuracy implicit in the position that government should do nothing. In not acting but in foreclosing others from a last resort to necessary physical force, government takes a position giving clear advantage to those who can realize their ends without force. In this sense, government cannot "do nothing." On the government monopoly of force, see Hurst (1960).

2. One case mentioned by Coase (1960: 11-13), Bryant v. Lefever, 4 CPD 172 (1878-79), illustrates beautifully the reciprocal nature of the social-cost problem. In this case, the plaintiff and the defendants occupied adjoining houses. Before 1876, the plaintiff could light fires in his house without the chimney smoking. But in that year, the defendants rebuilt their house to a new height, cutting off the

circulation of air over the plaintiff's chimney and causing his house to fill with smoke whenever the plaintiff lit fires. The trial court found the defendants' activities had created a nuisance and awarded plaintiff damages. The appellate court reversed, holding that plaintiff's activity in lighting fires caused the nuisance. Coase wrote of these opinions:

> Who caused the smoke nuisance? The answer seems fairly clear. The smoke nuisance was caused both by the man who built the wall and by the man who lit the fires. Given the fires, there would have been no smoke nuisance without the wall.

3. Economics is often attacked for its unrealistic assumptions. See, e.g., Hamill, (1968: 280), where he wrote: "All economic theories make general assumptions about the economic behavior of individuals, firms and governments. These assumptions seldom are realistic." Hamill, of course, assumes that unrealistic assumptions are not helpful. While this is in some cases undoubtedly true, in other cases it clearly is not. The unrealistic assumptions in Coase's model, for example, have great utility. They help identify the ways in which the real world diverges from the economist's ideal model, thus, in turn, they help to identify targets for government action designed to change the world (if that is what we wish to do).

4. For example, Louis Fuller, former head of the Los Angeles Air Pollution Control District, said: "No cost concerns me. As far as health and well being are concerned, no cost;" according to a transcript of The Advocates, KCET Television, Los Angeles, October 5, 1969.

5. See generally "Air Quality Act of 1967," 42 U.S.C., Sections 1857-57e, Supp. V, 1970; and Middleton, (1968). The Air Quality Act was amended by the Clean Air Amendments of 1970, Public Law No. 91-604 (December 31, 1970). The amendments made some changes in the pattern described in the text; for example, federal air-quality standards and some federal emission standards are now provided for. But the amendments do not make obsolete the questions of process described in the text; indeed, they simply raise them in a new context.

6. See also Esposito, Vanishing Air, 1970: 280-87.

7. See "Air Quality Act of 1967," Sect. 108 c 1, 42 U.S.C., Sect. 1857d, c 1 Supp. V, 1970. The hearings may in some instances have resulted in more actual participatory democracy than the draftsmen of the Air Quality Act anticipated. See items from an account of events in Pennsylvania, reported in the Wall Street Journal, October 21, 1969,

pg. 1. Since passage of the 1970 amendments, standard setting is now a federal job, and public hearings are not provided for. However, the states must hold public hearings before adopting implementation plans. See "Clean Air Amendments of 1970," Public Law No. 91-604, Sect. 4a, 1970.

8. For a valuable interdisciplinary study, one that was undertaken by students and that documented the extent to which the public could be excluded from participation in the air-quality decision process and made constructive suggestions for change, see "Air Pollution in the San Francisco Bay Area" (1970, Pt. 3). For other examples of the sorts of product which interdisciplinary research in the area of air pollution decisionmaking could yield, see the article by Hagevik, a planner (1968), and the comments thereon by a lawyer, Mandelker (1970). For an example of the sort of rigorous and valuable interdisciplinary research to which law-trained people can contribute, see DeVany, Eckert, Meyers, O'Hara, and Scott (1969).

9. See, e.g., Yannacone (1969: 14), where he noted: "The basic priority is recognition by our courts that the public has a right to a salubrious environment."

10. See e.g., the interesting paper by Murphy (1969), which suggests many of the specific sorts of questions into which lawyers may fruitfully begin inquiry. For a constructive inquiry into specifics-- into the role a specific legal institution (the courts) can play in regard to a specific problem (that of democratizing the process of decisionmaking about environmental quality) through a specific legal doctrine (that of public trust) see Sax ("Public Trust," 1970).

11. Of course, many people would simply take the position that Coase's theorem is irrelevant to the liability decision in any event, that pollution is and should be forbidden or penalized on purely moral grounds. In principle, there is nothing wrong with this position, so long as one recognizes that he may be sacrificing the gains of an efficient resource allocation for the gains of moral satisfaction. In practice, the issue may turn on discovering the basis for a moral judgment that pollution is bad, or at least that it is morally worse than some alternatives to pollution activity, such as decreasing industrial production at the cost of fewer jobs for poor people.

12. I owe these observations and the assembly-firm example to Michael E. Levine of the University of Southern California Law Center. For an example of the sorts of contrariness which can be discovered in assignments of liability if one is sensitive to other than the traditional and moral considerations, see Atwood (1969: 298). Both Atwood and Wright (1969: 383-98) take issue with portions of Coase's analysis.

13. But this is not universally true. See Renken v. Harvey Aluminum, Inc., 226 F. Supp. 169 (D. Ore. 1963).

14. The Clean Air Amendments of 1970, Public Law No. 91-604, Sect. 12a, 1970, provides for civil actions by citizens against the Administrator of the Environmental Protection Agency (charged with the federal enforcement responsibility of federal air pollution legislation) "where there is alleged a failure of the Administrator to perform any act or duty under this Act which is not discretionary with the Administrator" [emphasis added]. Query the extent to which this provision eases the citizen's problems.

15. See also Michigan's Environmental Protection Act, Public Act, No. 127, 1970, Mich. Stat. Ann., Sect. 14.528, 201-07.

16. For valuable discussions tracing judicial developments in the law of standing up to the time of this study, see Hanks and Hanks (1970); Sax ("Public Rights," 1970).

17. For a suggestion that the public-trust doctrine could be expanded into an effective instrument for judicial review of agency planning and enforcement decisions, see Sax ("Public Trust," 1970).

18. See also Michigan's Environmental Protection Act, loc. cit.

19. Cahn, "Environmentalists Blaze Legal Trail to Preserve Nature," Christian Science Monitor, October 2, 1969, p. 3, col. 1. See also Yanngcone (1969); Carter, both entries (1969).

20. See discussion of legislative facts by Davis (1955: 952-53). Legislative facts are those that bear on the court's job of creating law and policy, as opposed to adjudicative facts that define the circumstances of the parties to the litigation. For example, whether marijuana causes broad harm to public interests might be a legislative fact (Can the legislature constitutionally outlaw marijuana?); whether the defendant was smoking it or something else would be an adjudicative fact. Legislative facts are not subject to some of the usual evidentiary rules.

21. These papers and the discussion in the text focus on proof problems in the context of private suits for damages or injunctive relief. Proof problems can also arise, of course, in the process of participation in administrative proceedings and suits seeking judicial review of them. On some promising developments with respect to the burden of developing the record in administrative proceedings, see Hanks and Hanks (1970).

22. See, e.g., Reynolds Metals Co. v. Yturbide, 258 F.2d 321 (9th Cir.), cert. denied, 358 U.S. 840 (1958). See, generally, Prosser (1964: 218); Kimball (1969).

23. American Law Institute (1965: 26-27 Sect. 286). Statutory provisions have been accepted by the courts as a basis for civil liability in actions based on theories other than negligence, such as trespass, nuisance, and even strict liability.

24. On the problems raised by multiple causes and the judicial responses to them, see, generally, Harper and James (1956, Vol. 2, Sect. 20.3); American Law Institute (1965, Sect. 433 B(3)). See also Katz (1969: 616-21). On the question whether multiple polluting activities might give rise to liability even though the acts of each individual polluter are independent and, considered alone, would not be tortious, see Prosser (1964: 257). For a discussion of the bearing of federal air-quality criteria documents on proof of injury, see Borchers and Miller (1970).

25. For a more thorough discussion, see Baxter (1968: 44-53); as well as Hines (1966: 200-1), listing the fortuity of litigation, lack of judicial expertise and administrative capability, and nonrepresentation of the public interest.

26. For example, in Riter v. Keokuk Electro Metals Company, 248 Iowa 710, 82 N.W. 2d 151 (1957), plaintiff filed a pollution class action which was never joined in by any potential class members. But see Conti, "Conservationists Press Suits to Assert Right to a Clean Environment," Wall Street Journal, March 26, 1970 (Pacific Coast ed.), p. 1, col. 6, report on recent filings of pollution class actions: "There's ample evidence that the suits filed so far are just a start."

27. McElwain v. Georgia-Pacific, 245 Ore. 247, 421 P.2d 957 (1967). Contra, Fairview Farms, Inc. v. Reynolds Metals Co., 176 F. Supp. 178 (D. Ore. 1959).

28. Borchers and Miller (1970); it has been suggested that the right of individuals to enforce the federal pollution laws in private suits is ripe for judicial recognition. See Currie (1969: 23). The citizen suit provision of the Clean Air Amendments of 1970, Public Law No. 91-604, Section 304, makes such suits possible. Subject to certain notice provisions (or the commencement of a suit by state or federal officials), a citizen may bring a civil action against any person (including the United States and any other governmental "agency" to the extent permitted by the eleventh amendment), who is alleged to be in violation of applicable emission standards under the Act or orders

issued with respect to such standards by federal or state officials.

29. Diamond v. General Motors Corp., No. 947429 (Super. ct. Cal., Aug. 20, 1969).

30. On the polycentric nature of resource-allocation problems and their insusceptibility to satisfactory resolution through the process of adjudication, see Fuller (1960), and see also Nuestadter (1970). But see Committee on Pollution (1966: 231-34), suggesting that the judicial process as a means of alleviating the worst effects of pollution, and as a means of shifting some costs of pollution back to the polluter, can be made more efficient for these purposes by reforms that would do the following: First, establish an office of ombudsman to receive private complaints and sue, at the discretion of the ombudsman and at public expense, on behalf of private complainants. The various foundation-funded, public-interest law firms being established around the country could serve the same purpose to a limited extent. For a study of an agency that assumed the roles of environmental ombudsman and advocate and for a proposal that each role be formalized, see Krier (1971). See also H.R. 18242, 91st Cong., 2d Sess. (1970), proposing, among other things, an environmental ombudsman and an environmental legal-services office. Second, designate one or more courts in a jurisdiction to hear all pollution cases, in order to develop specialized judges. The courts should be authorized to appoint an appropriate pollution-control agency as a master in chancery to aid in technical aspects of litigation. The latter is a common practice with respect to water-rights adjudication and deserves careful consideration. And third, authorize the court, when appropriate, to broaden the area coverage of an action to include, if necessary, all polluters in a problem shed.

31. For a catalog of air pollution legislation on the state and local level, see Anonymous (a) (1967); Anonymous (1968). On the federal level, see, e.g., Fromson (1969); Muskie (1968).

32. We do, on the other hand, have a mass of descriptive studies of federal, state, and local regulatory programs in the legal literature. For the most part, these studies simply discuss the provisions of a particular program; at times, however, they engage in some valuable analysis and criticism. The following list is not all-inclusive:

For federal legislation, see Borchers (1970); Welch (1970); Edelman (1966 and 1968); Kennedy and Weeks (1968); Martin and Symington (1968); Middleton (1968); O'Fallon (1968); Reitze, "Role of the Region" (1969); Note (b) (1968); Johnson (1968); Note (c) (1968).

For state and local legislation, see Jurgensmeyer and Morse

(1968); Pollack (1968); Warren (1967); Ricksen (1958); Cowan (1955); Moran (1955); Chass and Feldman (1954); Kennedy (1954); German (1949); Zimney (1970); McCauley and Morgan (1969); Bell and Norvell (1969); Aborn and Axelrod (1968); Note (d) (1968); Note (e) (1968); Hamel (1965); Brestel (1962).

For general information, see Air Conservation Commission (1965); Edelman, "Air Pollution Abatement . . ." (1968); Schueneman (1968); Staff of the Senate Committee on Public Works (1963); Note (f) (1949).

33. See Haar (1959: 130); also see Schrenk (1948).

34. For illustrations of the point that enchantment with regulation is a real barrier to constructive change with respect to air pollution control, see "Letter to the Editor," from Seymour Schwartz (then a doctoral fellow, NAPCA, University of Southern California), Los Angeles Times, December 20, 1969, pt. 2, p. 4, col. 6; Hill, "Objections to a Tax on Pollution," New York Times, December 10, 1969, p. 33, col. 3.

35. For more detailed discussion, see Pollack (1968); Walker (1968); Note (e) (1968). Zoning, by itself or in conjunction with other regulatory techniques, has been used as a means of pollution control. In essence, of course, zoning puts the air pollution source "downwind." It has been suggested that to be effective, zoning constraints "should be couched in terms of emission royalties of license fees graduated by location rather than by black-and-white prohibitions here and permissions there," and should be combined with a program of property taxes designed to encourage optimum use of downwind and upwind land (Air Conservation Commission, 1965: 295); see also Dales (1968: 88-92).

Some (see Esposito, "What to Do," 1970: 48), have argued that zoning has "served to exacerbate pollution problems rather than to isolate residential areas from environmental fallout." The argument is based, first, on the point that "those who reside within the industrial area [are] often deemed to have foregone any nuisance claims," and, second, on the contention that "those who live outside the area are subjected to increasing amounts of pollution in the aggregate, but from increasingly numerous, remote, and unidentified sources." Residents within an industrial area might lose a nuisance claim either on the theory that they had come to the nuisance, i.e., moved into the area knowing of its conditions, or on the theory that no nuisance exists because the area has been devoted to activities which inevitably cause pollution; see Prosser (1964: 620-21, 623-33). In light of the fact that it is generally relatively poor people who "choose" to reside in industrial areas, the implications of either of these theories is painfully obvious. Compare Ross v. Butler, 19 N.J. Eq. 294, 306 (1868).

36. See e.g., California Health and Safety Code, Sections 39100, 39107, 39129 (West Supp. 1970); Clean Air Amendments of 1970, Public Law No. 91-604, Sections 6-8 (1970).

37. This is not to say, of course, that variance provisions in a performance program are not subject to similar dangers of abuse. See Air Pollution in the San Francisco Bay Area (1970: 226-30).

38. For example, Los Angeles, which essentially uses a specification system, has twice the sources of the San Francisco Bay Area Air Pollution Control District, which employs essentially a performance program. Los Angeles has 78 engineers, and the Bay Area 12; on the other hand, the Bay Area has 28 inspectors as compared to 71 in Los Angeles (about the same two-to-one ratio which exists as to sources). But see Note (e) (1968: 242), suggesting that a specification program requires fewer inspectors than a performance program.

39. Walker (1968: 84), suggests that this pattern results because performance standards are insufficiently precise to support criminal convictions without fear of sound constitutional attack. This, of course, does more to explain why a performance program might not use criminal sanctions than why specification programs ordinarily do. The real explanation probably lies in the forces behind the regulatory tradition, coupled with the fact that those forces are least restrained in the context of a specification program.

40. But see Hagevik (1968: 369), suggesting a regulatory scheme which would provide for bargaining between polluters and the control agency and would, it is argued, tend to produce optimal results in terms of resource allocation. For some arguments against Hagevik's plan, see Mandelker (1970: 209-10).

41. For example, in some circumstances, individuals injured by polluting activity can recover damages from a polluter even though the latter is in compliance with applicable regulations. Cf., Urie v. Franconia Paper Corp., 107 N.H. 131, 218 A.2d 360 (1966); what is authorized by law cannot be a public nuisance, but this does not affect any claim of a private citizen for damages for injuries caused by the authorized act but not experienced by the public at large. Assuming perfect compliance with regulatory standards and assuming further that perfect compliance produces the air quality desired by the legislative body, application of the rule stated in Urie would tend over

AIR POLLUTION AND LEGAL INSTITUTIONS

time to result in a higher level of quality than the legislative standard. One could thus argue that the function of the rule is to correct standards clearly below public desires, especially because its application would increase as more and more people became less and less satisfied with present standards and more willing to take on the burdens of lawsuits in which the rule might govern. Of course, no control program enjoys perfect compliance, nor is any likely to achieve precisely the desired level of air quality. Thus, the rule stated in Urie might be more understandably viewed as a way of filling gaps in program design and administration, as well as a way of compensating those who suffer most directly and seriously from legitimate polluting activity--something a regulatory program does not achieve.

42. Except for certain taxpayers with prior commitments, the federal investment credit was repealed for purchasers after April 18, 1969. See the Internal Revenue Code of 1954, Section 49; on accelerated depreciation, see Section 169.

43. For a convincing argument that prices should not be lower in remote, thinly populated areas, see Dales (1968: 88-93). Dales argues that, to accept the position that of two factories producing the same amount of emissions, the one which does less damage (because, for example, it is located in a barren desert) should pay less taxes would tend to push polluters into lower-tax areas. This in turn would tend to even out pollution geographically and make levels the same everywhere. He believes that people want separate facilities--areas of high concentration and areas that are virtually pollution free. He fears that with pollution levels the same everywhere, no area will be good; he believes there should be good and bad areas, among which people could choose, based on their individual assessments of advantages and disadvantages of each area (e.g., lower v. higher pollution levels, lower v. higher land costs, lower v. higher wages). Accordingly, he argues that the authorities should adopt either a polluter, locationally neutral, pricing policy (i.e., that they should set equal prices throughout a region), or a policy of relatively lower prices within highly polluted areas of a region. He argues that, as a result of either policy, business would tend to concentrate in highly polluted locations rather than spread to clean areas.

Pricing is said to maximize the incentive to discover technological improvements. As we have seen, subsidies can create the same incentive, but only in a less efficient manner (higher payment costs or higher administrative costs). A performance-regulatory system can also create the incentive, but only on a discontinuous basis; the incentive ceases once the imposed emission standard is met. To surmount this discontinuity would entail tremendous administrative costs.

44. E.g., according to Hagevik (1968: 373), the program would "have to survive legal attacks based on arguments of apparent discrimination and abuse of taxing power." It has been suggested that the Federal Air Quality Act of 1967 contemplates only direct regulation (Havighurst, 1968). The language of the Act, however, does not appear to compel this conclusion. See Air Quality Act of 1967, Section 108c 1, 42 U.S.C., Section 1857c $\overline{1,}$ Supp. V, 1970. The same can be said of the 1970 amendments; see Clean Air Amendments of 1970, Public Law No. 91-604, Section 4a.

45. It is fascinating to consider the background role to be played by the courts if a pricing system of control were to be adopted. Under a regulatory regime, parties injured by a polluter's activities always have a case for the recovery of damages (on such theories as nuisance, trespass, strict liability, negligence per se), if the polluter has violated applicable control regulations and may have a case even if the polluter has not. (See note 41.) Should the same results hold under any of the possible pricing models? The problem seems quite simple in the case of the auction model, as long as the polluter has exceeded the pollution rights he has purchased. But suppose he has not? In the case of an emission fee, should the polluter ever be required to pay damages in a lawsuit as long as he has paid the appropriate tax? If he is required to pay, should he be given a credit against his tax in the amount of the damages? Or should the problem simply be resolved as it has been under a regulatory regime, and for the same policy reasons?

46. For enlightening comments on research, design, and implementation, as well as on areas of needed research, see Herfindahl and Kneese (1965: 81-96).

47. Some examples are the Council on Environmental Quality, created by the National Environmental Policy Act of 1969, 83 Stat., 852, and California's Environmental Quality Study Council, Cal. Govt. C., Sections 16000-80 (West, 1970). See, generally, Muskie (1970).

48. See Ruff (1970). On political problems confronted in programs of regional control, see Lieber (1968). For interesting discussions of a regional air pollution control program, see Maga (1965); and Haydel (1967: 702-11).

49. See, e.g., The Environmental Protection Agency (EPA), established by Reorganization Plan No. 3 of 1970, July 9, 1970. EPA has been assigned many of the duties of setting and enforcing environmental-protection standards previously scattered among disparate federal agencies. Similar agencies exist in several states. See, e.g., N.Y. Environ. Conserv., Sections 1-129 (McKinney, 1970); Ore. Rev. Stat., Sect. 449.001-.855, 1970.

CHAPTER

6

**THE ECONOMICS
OF AIR
POLLUTION:
A LITERATURE
ASSESSMENT**

Robert J. Anderson Jr.
Thomas D. Crocker

ECONOMIC THEORY AND AIR POLLUTION

It now seems well established that pollution in general and air pollution in particular is the most broadly based political issue of this and perhaps any other time. Virtually all segments of the body politic manifest concern that environmental resources be shepherded more wisely. The media bristle with predictions of the dire consequence of unchanged ways. Part of this attention may simply be a manifestation of that faddish concern with natural resource and conservation issues which inexplicably wells up periodically throughout the developed world.[1] In substantial measure, however, today's preoccupation appears to be attributable to economic development itself. This proposition seems to have acquired some currency in the literature, its Marxist overtones of economic determinism notwithstanding.

One of the more frequently conjectured economic causes of an aroused populace is the hypothesis that environmental quality (a pleasing environment), including clean air, is a strongly superior good (Crocker, Natural Resources, Journal, 1968; Gaffney, 1966; Krutilla, 1967). That is, as real, personal incomes increase, the demand for a pleasing environment increases in greater proportion than the increase in income. It true, the hypothesis suggests that continued, increasing, personal-income levels are likely to increase the public's excess demand for improvements in air quality even if we assume that present air-quality levels do not change a great deal.

Of course, on the basis of past experience, it is indeed an unlikely assumption that environmental quality will remain unchanged

as economic growth proceeds. If anything, many would argue that
what has been a minor ecological aberration in low population and
economically primitive settings becomes an ecological trauma of
major significance under sustained economic development (Erlich,
1969; Reinow and Reinow, 1969). Because all materials and energy
removed from the natural environment must ultimately return, the
increased usages associated with larger populations--each element
of which consumes increasing amounts--imply that environmental
loadings of the residuals from production and consumption must
increase. One result is air pollution. As development progresses,
the impact of these increased loadings is accentuated by man's apparent tendency to abandon a more or less independent and dispersed
living pattern to crowd together in cities. This crowding too often
serves only to place severe stress on an environment whose assimilative capacity is already strained.[2] Even in a static state, since matter
and energy are unavoidably lost in the economic system, maintenance
of a constant population at a steady level of economic well being
requires, in the absence of technical change, continual displacement
of matter and energy from natural sources and concomitant generation
of at least some residuals.

In sum, as development progresses, at least development as it
has been historically manifested, forces are typically unleashed that
cause demands for a pleasing environment to be intensified while, at
the same time, environmental quality is unrelentingly degraded. It is
the confluence of rising demand with increasing unavailability that
seems to be, at base, the source of today's anti-pollution fervor.
Allocation systems have not evolved in a manner permitting the
reconciliation of these tendencies.

Indivisibility and the Market

It is only honest to note that the pollution problem is somewhat
of an embarrassment to the economics profession. Nearly two hundred
years of professional effort have been devoted to progressive refinenent
of the proposition that the market works at least tolerably well in
allocating resources. The apogee of achievement has been the demonstration that, under certain conditions on the character of the economic
system, a unique competive equilibrium exists which is also Pareto
efficient.[3] Although for some time, casual attempts at empirical
verification of the hypotheses of competitive equilibrium welfare
theorems have suggested that the hypotheses are not in fact fulfilled,
there has persisted the blind faith that even major deviations from
the theoretical preconditions of Pareto-efficient, competitive equilibrium still produce market outcomes which approximate the Pareto-efficient ideal. However, it is now becoming painfully apparent that

what have heretofore been thought to be minor welfare effects in isolated parts of the economic system are really neither minor nor isolated. The air resource allocation problem, as manifested in the current, widening disparity between realized and apparently desired levels of air pollution, is but one example of many ostensible welfare failures occasioned by the market's present configuration.

In particular, because the ambient air must be the same at more or less contiguous locations, air quality is very much a joint good (Oakland, 1969). Whatever quality one shall have, all will have, but the individual's marginal rate of substitution among alternative uses need not be unity. To better understand the consequences of indivisibility, consider a simple, spatially compact economy with r commodities and factors and N consumers. Let the nth commodity be ambient-air quality and let

$$U^i = U^i(q_1^i \ldots q_r^i)$$

be the utility of the ith individual. Further, let the social-transformation function by given by

$$F(q_1, \ldots, q_r) = 0$$

The technical indivisibility of the air resource imposes the constraint that all must enjoy (or not enjoy) the same ambient-air quality,

$$q_n^1 = q_n^2 = \ldots = q_n^N = \bar{q}_n$$

where $\bar{q}_n$ is any arbitrary, prescribed, ambient-air quality. If we assume that certain differentiability and curvature conditions are fulfilled by the individual-utility functions and the social-transformation function of the revelant economy, the first-order conditions for a Pareto-efficient allocation under such a regime are given by

$$\frac{U_j^1}{U_k^1} = \frac{U_j^2}{U_k^2} = \frac{U_j^n}{U_k^n} = \frac{F_j}{F_k} \qquad j = 1, \ldots, n-1, n+1, \ldots, r \qquad (1)$$

$$\sum_{i=1}^{N} \frac{U_n^i}{U_k^i} = \frac{F_n}{F_k} \qquad k = 1, \ldots, n-1, n+1, \ldots, r \qquad (2)$$

where the subscripts indicate partial differentiation, e.g.,

$$U^i_j = \partial U^i / \partial q^i_j \text{ and } F_j = \partial F / \partial q_j.$$

As is well known, there are in fact an infinite number of distributions of $(q_1, \ldots, q_r)$ which are consistent with the fulfillment of the necessary conditions.[4]

It is a standard classroom exercise to write down the utility and production functions for all economic units in some blackboard society and demonstrate how, in the absence of externalities, the system of equations derived from the maximizing behavior of these units may be reduced to a set of equations of the form of 1 above. But when a single market price for air quality obtains, the conditions in 2, above, cannot be synthesized in like fashion out of decentralized maximizing behavior. Only some extraordinarily complicated, multipriced system would be consistent with decentralized attainment of these conditions.

One factor that contributes to the prevention of such attainment is the difficulty of excluding individuals from being exposed to whatever air quality happens to exist in their physical environs. Each receptor or sufferer from the disamenities of a polluted atmosphere realizes that he can benefit from the decision of any other, similarly motivated unit to "bid away" the air resource from those who use it as a vehicle for waste disposal. The result is usually an impasse in which each receptor hopes the others will act, while a smaller than optimal quantity of clean air is provided because emitters usually face a near-zero price for their waste discharges. Emitters are encouraged to make ever more intensive use of the air resource in an attempt to drive the marginal-revenue product of the air used as a waste-disposal medium to near zero.

In sum, then, it seems unlikely that the incentives which organize and maintain markets could bring about a set of prices consistent with the attainment of 1 and 2. However, it should be emphasized that there do exist some quite special situations in which rights to the use of the air resource can be traded in a market which functions in order to attain a Pareto-efficient outcome (Coase, 1960).[5]

The Central Role of Extramarket Institutions

Implicit in the argument to this point has been the assumption that the responsibility for initiating bargaining over the air resource lies with the receptor. In point of fact, this appears to be the case under current law. It should be explicitly noted, however, that this

THE ECONOMICS OF AIR POLLUTION					137

is a consequence of the institutions man has established; it is not in
any fashion a result of some immutable, physical reality. Although
the law apparently awards de jure rights to the superadjacent
air resource to the property owner, the statutes under which proceedings may currently be brought are distinctly unfavorable to the receptor. Thus, de facto, emitters may do pretty much as they please.
Under such conditions, there is no incentive for emitters to initiate
bargaining for a resource available free of charge. Due to the mobile
and fugitive nature of the ambient air, this de facto institutional
arrangement exacerbates what might otherwise be a resource-allocation problem amenable to a market solution.

Even apart from the problems occasioned by technical indivisibility of the air resource, there are reasons for supposing that de
facto assignment of property rights to emitters could well result in a
different air quality than would assignment to receptors. First, the
informational, contractual, and policing (ICP) costs of exercising
rights to the ambient-air resource often seem to be asymmetric
between emitters and receptors.[6] In general, ICP costs for receptors
are likely to be greater than those for emitters. There are several
reasons for this: Receptor uncertainty about each emitter's contribution to the pollutant dosages to which he is subjected, the instrumentation, time, and personnel necessary to monitor stack gases, and
the atmosphere's pollutant content coupled with an immense number
of receptors relative to (major) emitters all conspire to suggest
that where the burden of proof is placed on the receptor--whatever
be the de jure distribution of rights--the emitter is likely to enjoy
the air resource's use virtually unfettered.

However, differential ICP costs are not the sole causes of the
differences in bargaining or market outcomes caused by alternative
property rights assignments.[7] These assignments can also have
demonstrable income effects, because alternative assignments imply
alternative distributions of producer's and consumer's surpluses
(Dolbear, 1967). In general, assignment of rights to emitters would
result in poorer air quality than would assignment of rights to receptors (Mishan, "Pareto Optimality," 1967).

It is important to keep in mind that the concept of Pareto efficiency is inherently a constrained concept. At one level, this is
obvious. For years, economists have been finding solutions to
qualitative problems of maximizing welfare subject to constraints on
resource quantities. Recently, however, economists have learned
that there are important institutional constraints which, for the sake
of realism, ought also to be formally incorporated into their heuristic,
constrained, maximization problems. Recognition of the importance
of such constraints means that it is no longer possible to be secure

in the conclusion that market failure is the unavoidable consequence of some inviolable physical principle or principles. It is no less likely that market failure is a direct result of well-known human fecklessness in structuring social affairs to produce an outcome which is in some sense optimal.

The Potential Role of Economics in Air Pollution Control

If, by altering the constraints under which exchange processes operate, it is possible to change the set of feasible outcomes, the problem of air pollution control, in the most fundamental sense, is to discover that set of manipulable constraints which, for a given state of those constraints not subject to control, will achieve and maintain a desirable allocation of the air resource. That is, the air pollution control problem consists of finding an arrangment that elicits emitter and receptor behavior minimizing the sum of the social costs associated with the damages due to pollution, the control of pollution, and the resources used by market and nonmarket institutions to promote air pollution control. Discovery of such arrangements requires knowledge about the value of a large number of natural and economic parameters and the effects thereon of changes in institutional constraints. For the most part, such knowledge is lacking. Air pollution control must presently proceed under substantial uncertainty.

In effect, the typical response to this uncertainty has been the lodging of all rights in the air resource to a central-control authority whose responsibility it is to allocate the air resource among emitters and receptors. To date, the criteria upon which such allocation decisions have been based appear not to have included economic efficiency. Instead, decisions on environmental-quality objectives have seemingly been based on a strictly lexicographic ordering of alternative environmental states.[8] The highest rank in the ordering is consistently accorded the prevention of health impairment of even the most sensitive elements of the population. Pollution loadings that exceed levels at which impairment of health has been noted are simply held to be unacceptable.[9] Decisions taken with respect to attaining the selected environmental objectives are too often byproducts of comic political intrigue and a misplaced concern for treating emitters equitably.

If it happens that health effects are observed only at relatively high dosages, more flexibility in the decision criteria is sometimes witnessed as consideration is given to property damages and aesthetic impairment. In general, however, while debate surrounding air resource allocation decisions is frequently peppered with preposterous emotional claims and counterclaims about the economic consequences that will ensue if a given policy is adopted, singularly little attention appears to have been accorded to the small body of scientific economic knowledge available. For example, under the abatement strategy recommended by the Secretary of HEW at the conclusion of an abatement conference in the Washington, D.C., metropolitan area, it would cost about one hundred times as much to achieve the recommended air-quality goal as it would cost under an alternative strategy achieving the same goal (Ernst and Ernst, "Cost-Effectiveness" 1969). Although considerations other than purely pecuniary efficiency are admittedly highly relevant to the implementation of air pollution control, it is difficult to believe there was no acceptable middle-ground strategy less expensive than that recommended.

The ultimate social worth of research hinges critically, of course, on its usefulness. If, as seems to have been the case in the past, economic research results are ignored by those responsible for institutional reform and/or air quality management decision ("Don't confuse me with the facts; my mind is already made up"), it is difficult to make a case for expanded, economic-research efforts. It is also legitimate to ask: "How do the social costs and benefits of a control strategy that includes an economic-research program compare with one which does not?" We shall return to these points in the concluding section of this chapter.

A Reader's Guide and a Caveat

We now turn to a review of past work on the economics of air pollution. This specialty field is so new that no taxonomy has been proposed for its various constituents. For want of an established classificatory scheme, we have catalogued contributions to the literature under the following headings:

Theoretical and empirical studies of the emitter
Theoretical and empirical studies of the recptor
Theoretical and empirical general equilibrium models
Theoretical and empirical studies of institutional
 instruments of control.

We are inexperienced bibliographers and therefore caution the reader not to dimiss a reference we have discussed under a heading which seems unrelated to his principal interest. It should also be kept in mind that the field of air pollution economics is so young and the unknown so vast that it is wise to be critical of our assessment of the merit of the work reported in the various references. Prudence in this case, as always, demands that the reader maintain scholarly skepticism.

THEORETICAL AND EMPIRICAL STUDIES OF THE EMITTER

The "theory" of the emitter is essentially an application of the theories of the household and the firm, concentrating in particular on the waste disposal practices of these decision units. As with traditional economic decisions typically ascribed to these units, emission decisions are assumed to be made in order to maximize profits (in the case of firms), and utility (in the case of households). From the emitter's point of view, the economic problem is one of minimizing all costs, including waste-disposal costs, for any given level of production or utility obtained. Thus, the fundamental question for the emitter is the cost of methods of production and consumption that are contingent on discharge of wastes into the ambient air relative to the costs associated with methods dependent on other methods of disposal.

Existing empirical work on emitter control cost functions falls into three subcategories: (a) engineering studies, (b) survey studies, and (c) combinations of the two.

Engineering Studies

Engineering studies have traditionally focused on capital costs of increased tail-end control, preproduction modification of inputs to reduce the pollution-generating potential thereof, and process modification. Operating costs usually receive considerably less emphasis, especially in studies of tail-end control. Costs are usually expressed on an annual basis and include a somewhat arbitrary charge for interest, depreciation, and whatever (if any) allowance the investigator wishes to make for operation. Allocation of joint costs is similarly ambiguous. Basic cost data, for the most part, seem to come from manufactures' list prices and the investigators' so-called average experience with actual operating units. Rarely is any attempt made to account for factors other than engineering-design variables that may account for cost variations.[10]

Engineering studies comprise by far the greater part of the control-cost literature, so it is impossible to give explicit mention to more than a few papers here. There are several representative examples of tail-end control cost studies (Semrau, 1963; Katell and Plants, 1967; NAPCA, "Control Technology," 1970: 6.1-6.6). Nearly all studies of conventional-power generation and automobile-emissions control, e.g., Faith (1966), and Jensen (1964), follow this general format. The issues of major interest in these studies and the many others like them are the relationships between collection efficiency, waste-material attributes, and physical configuration of the tail-end control equipment.

Engineering studies of input modification are best exemplified by investigations of the economics of fossil-fuel desulfurization (Bechtel Corporation, 1964; Frankel, 1968), and substitution of fuels with low sulfur and ash content for fuels with greater amounts of these elements (Frankel, 1969).

Studies of process modification as an alternative to input modification tail-end or exhaust control are not particularly common although investigations into costs of power sources for automobiles other than the internal-combustion engine, e.g., NAPCA (Prospects for Electric Vechicles, 1969), and of changes in stack height and other exhaust-gas properties affecting atmospheric dispersion (Nelson and Shenfeld, 1965), can be cited as examples.

The principal defect in the engineering approach to emission control cost-function determination lies not in the results obtained but rather in the interpretations thereof. Engineering studies are inherently single-technology oriented. The raison d'etre of these investigations is the determination of the costs associated with some particular method of emission control. As such, they do not answer the more general question on the actual costs of controlling emissions, even if one assumes that emitters act in order to minimize costs. Few if any substitution possibilities are considered in any one study. The approach thus conveys no explicit information on emitter-cost functions unless the technology for which cost estimates have been prepared is the only one available. This is not to argue that engineering investigations are without merit. Indeed, information on costs of applying specific control technologies can be of the greatest use in determining control-cost functions. Still, the preponderance of literature on this type of effort relative to that based on the other two approaches to be discussed suggests that the final research effort giving economic meaning to these engineering results remains to be undertaken.[11]

Survey Studies

The survey approach to emitter control cost determination remedies, at least in concept, the "relevance" defect of engineering studies. Emitters with common technological and economic characteristics have been surveyed to obtain data on those production and consumption processes that either generate or control emissions. Cost functions are then fitted to the data using statistical methods. (See Hamburg, 1969; Frankel, 1969; Woodcock and Barrett, 1970; and

O'Connor and Citarella, 1970.) In theory, because emitters choose least-cost methods, cost functions so determined presumably reflect consideration of relevant substitution possibilities. It should be noted, however, that survey information is usually extremely poor. Sometimes emitters are asked to make difficult allocations of joint costs, sometimes they are ignorant of costs associated with activities wholly directed toward emission control, and sometimes responses are purposely biased due to recognition that study conclusions may be employed for regulatory purposes (Jackson, 1969; and Fogel, 1970).

Engineering-Survey Studies

Finally, a third approach seeks to synthesize survey data with engineering information like that underlying the first approach. To our knowledge, the sole example of this approach is described in full by Anderson (1969), and is summarized in Anderson ("Application," 1970). In broadest terms, engineering theory is used to build a mathematical model of the plant; as completely as possible, it describes known methods for transforming inputs into outputs, where the outputs are basic products and pollution. This representation of production relations is the basis for formulation of the constrained cost minimization problem of the plant manager, namely, minimization of cost subject to constraints on output and on maximum allowable emissions. The resulting programming problem is solved for many different constraints on output and emissions and for many different combinations of factor prices. The solutions to these problems can convey a considerable amount of prior information about the character of plant costs and their dependence on air pollution control decisions. Furthermore, the use of such prior information materially aids in the interpretation of what is frequently very poor survey data.

Some Neglected Aspects of Emitter Behavior

While the current state of knowledge about emitter-control costs cannot be described as particularly impressive, even less is known about other equally important facets of emitter behavior. Though it is trivial to point out that emitter behavior is likely to be influenced by market structure as well as by technical alternatives,[12] it seems that studies which attempt to assess the impact of market structure on emitter behavior are practically nonexistent.[13] Consider the following issues to which no one has apparently devoted any attention.[14]

1. Highly competitive industries may respond quite differently than concentrated industries to particular control policies. Industries

THE ECONOMICS OF AIR POLLUTION

with some history of collusion among member firms may be better able to innovate in waste production and control technology because of scale economies and easier coordination of complementary research efforts. Or innovation could be stifled as is alleged by the U.S. Department of Justice to have happened in the automobile industry. What are the determinants of the rate of invention and innovation of emission-control technology?

2. What determines what a business is willing and/or able to spend on largely "nonproductive" undertakings such as pollution control? To the extent that control efforts employ a large measure of purely moral suasion, as appears to be the case today, such a question is extremely important.

3. How might emitters respond to various tax-subsidy incentives or cost-sharing approaches to encourage air pollution control?

4. What are the economics of recycling resources?

5. What effect will the imposition of air pollution control have on the structure of the firm and the industry?

6. What about the development of human capital needed by emitters to implement pollution control?

7. What new industries, e.g., firms specializing in the removal of wastes and consulting-engineering firms, are likely to result from pollution control, and what will the economics of these industries be?

8. Who will bear the burden of control? How much will retained earnings be decreased, price increased, output reduced, factor demands reduced, and so forth?

9. What would be the effect on price and output, e.g., the sulfur market or sulfuric acid market, if large quantities of exhaust gas were converted into these chemicals and offered for sale on the market? What would be the effect in similar instances where large amounts of some marketable product are recoverable?

THEORETICAL AND EMPIRICAL STUDIES OF THE RECEPTOR

In general, studies of receptors concentrate on defining and measuring net economic losses due to air pollution, i.e., residual losses plus the costs of adjusting to the presence of air pollution. In a world of certainty or certainty equivalence, residual damages for the individual receptor and for the collection of receptors consist of the difference in the value of production and/or consumption under the input proportions and scales prevailing before the advent of air pollution and the input proportions and scales prevailing after all adjustments to the presence of air pollution. Costs of adjustment are the value of the resources expended in moving along an optimal path from one equilibrium position to another (Crocker, 1968: Some Economic Aspects . . . 113-64).

In a world of receptor uncertainty about future air-pollution dosages, additional sources of air-pollution damages are brought to bear on the receptor. If the receptor is risk averse, so that he attaches disutility to an increasing spread of possible gains or losses, then his inability to prevent discrepancies between expected and realized outcomes has a cost. This cost is no less real than damages to his health, for he would be willing to pay to be rid of both the health damages and the uncertainty (Crocker and Rogers, 1971).[15] The receptor can alleviate the impact of this uncertainty somewhat by substituting relatively pollution-insusceptible inputs for pollution-susceptible inputs. His problem is to select that input mix which minimizes the losses he suffers due to discrepancies between expected and realized dosages of air pollution. Generally, the attainment of this flexibility requires the sacrifice of a higher output level over some relatively short time interval in order to ensure a high probability of achieving some lesser output level over each of several similar time intervals (Crocker, Some Economic Aspects . . . 1968: 316-21).

The presence of receptor uncertainty about future air pollution dosages generates another cost if receptor expectations are dissimilar (Crocker, Some Economic Aspects . . . , 1968: 316-21). It is one of the best known propositions of modern welfare economics that discrepancies in technical or subjective marginal rates of substitution among economic units cause the value of production to be below its potential maximum (de Graaf, 1967).

There exist four general modes of discovering air-pollution damages: (a) political, (b) the use of interviews with air-pollution sufferers, (c) the direct use of production and consumption response surface parameters, and (d) the use of market studies (Crocker, 1969).

Political Studies

Although perhaps most frequently employed at present, assessments of the intensities of political expressions, representations, and exhortations are not likely to be accurate indicators of what receptor preferences for one state of air quality relative to another would be in any real market situation. The intensities registered by political means represent only the relative valuations that occur under the constraints imposed by the political process. As such, they reflect the ability of the receptor to alter the relative prices he faces as well as the dependence of the receptor's optimal strategies on the difficult-to-predict actions of other participants (Buchanan and Tullock, 1962; Downs, 1957). No formal efforts appear to have been made to specify the magnitudes of receptor damages that usually emerge in these processes, although there do exist descriptions of the sequence of events in particular air pollution episodes (Goldner, 1966).

Interview Studies

The interview approach to the measurement of air pollution damages has problems similar to those of the political approach. In particular, unless the interviewee's statements of his willingness to pay for air quality are made in a context where the cost of imporving air quality is an integral part of his decision, he will overstate the intensity of his preference (Crocker, "Some Economic Aspects . . . " 1968: 241). Nevertheless, a number of studies have been made employing this approach (Medelia, 1965; Public Administration and Metropolitan Affairs Program, 1965; Loveridge, 1969; Strodtbeck, 1969). In spite of the free-rider problem that negates much of the validity of damage estimates obtained from the interview approach, the approach does have its usefulness. If there are pollutants whose damaging effects are extremely subtle so that they go completely unrecognized or are only recognized very slowly, the damage function can be changed by bringing these subtle effects to the receptors' attention. The interview assists policymaking bodies in acquiring an understanding of the degree of receptor comprehension of air pollution's effects. To the extent that this information is used to improve receptor comprehension, the use of the interview approach alters the damage function.

Response Surface Studies

The approach to measuring air pollution damages most frequently employed is estimation from production and consumption response surface parameters. Response surfaces are fundamentally analogous to production functions, a concept with which economists are more familiar. For any given state of factor influencing the behavior of some system (be it animate or inanimate), the response surfaces describe the output characteristics of the system. Applied to the problem of evaluating economic losses due to air pollution, the nub

of this approach is that known response surfaces are used to estimate
physical changes which would come about with changes in the levels
of pollutants in the ambient air. These physical changes are then
multiplied by the investigator's estimate of the changes' per-unit
values to obtain an estimate of the dollar significance of the response.
Examples of studies of this kind include the near infinitude of efforts
that have been made to translate the experimental and survey findings
about morbidity and mortality in organisms and deterioration of arti-
facts and materials into economic terms. The response-surface
approach has been employed in soiling studies (Michelson, 1966;
Ridker, Economic Costs, 1967: 57-89; Wilson and Minnotte, 1969;
Mellon Institute, 1913), health effects studies (Ridker, Economic Costs,
1967: 20-56; Lave and Seskin, 1969; Bates, 1967), studies of damages
to plants (Middleton, 1961; Hepting, 1964; Wolozin and Landau, 1966),
to animals (Cass, 1961; Harris, 1964), and to materials (Uhlig, 1950;
Armour Research, 1960).

The approach has several methodological problems, not all of
which are always satisfactorily resolved. Perhaps the most serious
methodological defect of damage estimates prepared in this manner
is that so little is known about the response surfaces on which they
are based. Where surface parameters are based on survey data, such
ignorance is an unavoidable consequence of limitations on the range
of sample experience. Variations in inputs are frequently small, and
one cannot be very confident that nuisance variables have been satisfactorily controlled. Where estimates of response surface parameters
are based on data from rigidly controlled laboratory experiments,
it is possible to entertain doubts that parameter values reflect empirically relevant portions of the surface. In particular, in order not to
confound responses to air pollution of the entity under study with other
stresses, input proportions and scales that yield near-maximum
marginal responses to pollution are typically employed. This suggests
that experimentally determined marginal response to an additional
dosage of pollution is likely to be relatively great compared with that
which would be empirically observed under ideal conditions in real
situations. In other words, simple extrapolation of experimental
results to real situations probably results in an overestimate of real
damages, if we assume that correct per-unit-dollar damage figures
are employed. It would be useful if the relationship between real and
experimental outcomes were rigorously investigated, for there do not
seem to be any careful investigations of this sort.[16]

Even if a conscious attempt were made closely to approximate
empirically observable conditions in the laboratory, the number of
conceivably relevant combinations of state variables is so large as
virtually to preclude a complete investigation. Any entity subjected
to air pollution dosages is likely to have available several physical

THE ECONOMICS OF AIR POLLUTION

substitution possibilities of varying degrees of susceptibility to the presence of air pollution. Each of these physical inputs has three temporal dimensions--rate of use, time of initial use, and time interval of use--that can be substituted for each other, and most of these temporal and physical substitution possibilities can be transported across space. For any given entity, the available physical, temporal, and spatial substitution possibilities to be accounted for is thus numerous. Of course, the investigator can ease the problem by combining prior qualitative knowledge about the technical features of these possibilities with a knowledge of their relative prices in order to rule out those possibilities that are uneconomic. However, we know of no cases in air pollution damage investigations using either the experimental or survey method where explicit attention was given to the relative prices of adjustment possibilities. In the absence of this explicit attention, it seems likely that a great many possibilities have been studied which have no economic relevance whatsoever.

It should also be noted that the response surface approach is primarily geared to the determination of damages suffered by single entities in isolation, whereas, from a policy standpoint, it is the aggregate of damages which are of primary interest. However, the collection of receptors cannot simply be treated as some arithmetic sum of individual receptors, for the prices of the substitution possibilities that the single receptor views as fixed are not necessarily fixed for the collection of receptors. The substitution of one input for another by a single receptor will usually not affect relative input prices, but if all receptors carry out the same substitution, relative input prices will often be affected. This problem is but one of many which make it exceedingly difficult to determine an appropriate per-unit-dollar damage figure to multiply by physical changes estimated from the response surface.

Other problems abound as well. For example, effects on human health are commonly considered to be the most serious economic losses due to air pollution. All studies of these economic losses concentrate exclusively on the loss of livelihood involved in human morbidity and mortality. Conceptually, at least, standard economic criteria can be employed to determine the economic losses caused by air pollution to buildings, crops, sites, and other forms of property. The interactions of the market determine a price for each characteristic of each good, and, for the individual, the replacement cost of a good sets an upper bound of the characteristic's value to its owner. The contingent nature of these damages can, in principle, be insured against. However, if air pollution damages the health of this owner or kills him, the economic loss he incurs is more difficult to ascertain because what his life is worth to him and his family is nontransferable. From the point of view of his family and himself, he is completely unique. No economic analysis, no matter how carefully done, will be able to account for all these factors. Nevertheless, it seems reasonable to assert that simple loss of livelihood or health care outlay measures of air pollution damages can be improved on, both conceptually and empirically.[17] For example, some insights might be gained by inquiring into the amounts individuals pay to take safety measures and purchase contingent claims in order to reduce the probability of losses due to sickness or death.[18]

Market Behavior Studies

Some of the problems associated with other approaches to damage measurement are circumvented by the fourth approach to the measurement of air pollution damages, the explicit use of market valuations. This approach explicitly considers the effect of dosages on market behavior.

With the use of census data or individual property sales records, a number of studies have applied this approach to the market for land and residential property. (See Ridker, Economic Costs, 1968: 141-51; Ridker and Henning, 1967; Zerbe, 1969; Anderson and Crocker, 1970; Crocker and Rogers, 1971; Crocker, Externalities . . . , 1971.) By the simple expedient of replacing distance variables with air pollution dosages in any one of a number of extant theoretical models yielding an inverse relation between land values and distance (e.g., Muth, 1961), an inverse relation can be obtained between land or property values and air pollution. In Strotz, ("The Use of Land Rent . . . ," 1967), given certain simplifying assumptions, it is shown that if two sites are similar in all respects except air quality, the difference in their value represents the market's willingness to pay for reductions in air pollution dosages. That is, all air pollution damages will be registered in the differential site values. However, if all consumers do not regard the two sites as perfect substitutes when each site has equal air pollution dosages, then some air pollution damages will be registered in other durable assets and losses in consumer's surplus. Nevertheless, it is shown that these sites and other assets enter individual utility functions in an additive manner. Thus, differential site values can be employed to obtain a lower bound on air pollution damages, and, if the sites in question have rather homogeneous characteristics, their differential values represent all or nearly all damages. Similar conclusions are obtained and strengthened in Lind (1970). Employing common multivariate-estimation techniques, these studies have universally found an inverse relation between property values and air pollution which, for residential properties, amounts to an average marginal capitalized loss of between $100 and $1,000

THE ECONOMICS OF AIR POLLUTION 149

per residential unit. One paper has attempted to employ court judgments of property-value reductions (Havighurst, 1970).

The advantage the market behavior approach provides, relative to other approaches to the measurement of air pollution damages, is ease and rapidity (Anderson and Crocker, 1969). The investigator does not have to discern and evaluate receptor-adjustment possibilities, nor does he have to worry about making individual transactions commensurate in order to aggregate them. The market does it for him through directly observable market prices. However, this approach, other than providing knowledge of the general form of the relation between net damages and air pollution dosages, provides little insight into the fundamental processes of adjustment whereby air pollution has its impact. What it does is describe the results of these processes of adjustment. It therefore has limited predictive content. Any thorough program of air pollution damage investigations must therefore consider the employment of all four approaches to measurement. The market-behavior approach can be used to obtain knowledge of existing damage magnitudes and the general form of the damage functions. It can thus be used to eliminate investigations by the technical-coefficients approach of uneconomic portions of the structure of adjustment possibilities. The interview approach can be employed to determine the effects, if any, that receptors fail to perceive.[19] Finally, the political approach permits an accounting of those effects that receptors perceive but which are not registered in the relative prices of adjustment possibilities or which control authorities are unable to perceive.

AGGREGATED SYSTEMS

A relatively recent development in the literature that promises to be of major use to the policymaker is the use of macroeconomic models to simulate the impact of alternative air pollution control strategies on regional and national economies (Teller and Norsworthy, "Regional Impact . . . ," 1969; Anderson, "A Note on . . . , " 1970; Ayres and Kneese, 1969; Rose, 1970). From the point of view of the policymaker or planner, endless concern over the specifics of emitter and receptor behavior are best kept to the researcher and technician. The policymaker is more likely to be interested in consistent summary aggregates of emissions, air-quality measures, employment, income distribution, consumption, and gross costs and benefits.

Work in this area may be divided into two categories according to the basic model employed--first, input-output models, and second, econometric forecasting models. The fundamental innovation in these new applications of old models is the explicit recognition of important mutual interdependence between the economy and the environment. Traditionally, economic theory has stressed the influence of the environment on the economy, e.g., Jevon's sunspot theory of business cycles and the emphasis on natural-resource endowments in some theories of trade and development. It has finally become evident, however, that the economy also profoundly influences the environment. The factor of production "land" is once again assuming importance in our analyses.

The work reported in Ayres and Kneese (1969), uses the input-output approach. An explicitly general equilibrium approach to assessment of interdependence between economy and environment is taken. Starting from the fundamental physical law that matter and energy may neither be created nor destroyed, gross wasteloads are projected on the basis of material and energy balances. Waste-control activities, it is recognized in the model, do not eliminate wastes, but only metamorphose them to a form that is presumably less detrimental to man and to the ecosystem than is uncontrolled waste. Allowance is made for the effects of accumulations of residuals on the operation of the economy. In addition to the great generality of their approach in a static context, these studies also note the possibilities generally neglected in theory of eliminating externalities by technical innovations rather than the standard tax-subsidy-compensation approaches. Work is currently being undertaken at Resources for the Future on development of heuristic numerical application of these models.

The results of Anderson, ("A Note on . . . ," 1970) and Rose (1970), are based on econometric forecasting models of macro-economics. Using a well known national model, Anderson (reported in Rose, 1970), forecast the impact on the U.S. economy of pollution control expenditures in amounts like those summarized in NAPCA, The Cost of Clean Air (1969). It should be noted that the forecasts so reported are extremely naive. There is no pretention that the simple changes made in the exogenous variables of the model adequately account for all structural economic changes which might be expected if the sweeping air pollution control program schematically envisioned in the calculation were actually effected. Rather, this forecast is best viewed as indicative of the potential of forecasting models to provide more comprehensive assays of the economic impact of air pollution control than have formerly been available.

The work by Anderson ("A Note on . . . ," 1970), is a first attempt at assessing the indirect effect of restrictions on the burning of high-sulfur coals in the economy of a region which is, in part, dependent on the mining and export to other regions of the United States of these high-sulfur coals.

THE ECONOMICS OF AIR POLLUTION

In 1971, CONSAD Research Corporation of Pittsburgh, Pennsylvania, was engaged in a longer term project to develop an interrelated set of regional forecasting models and a national model that would forecast both the effect of air pollution control policy on regional and national economies and the effect of a polluted environment thereon as well. All models were to be sectored in order to emphasize primarily those sectors that either influenced and/or were influenced by environmental quality to a significant degree. Up to 1971, models of thirty-one air-quality control regions had been built, and work was proceeding on (a) developing models for more air-quality control regions, (b) linking individual regional models with one another through import and export equations, (c) developing more detail in so-called air pollution sectors, and (d) linking the system of regional models with a national model.

CONTROL INSTRUMENTS

Importance of Property Rights

The efforts investigators have devoted to research on the technology of air pollution generation and control, and dosages and effects, seem consistent with the acknowledged truth that technical possibilities constrain the set of feasible outcomes. It is also widely acknowledged that individual receptor, emitter, and control-agency endowments constrain the set of feasible outcomes. These constraints serve to define the environment within which the individual must make choices. Not so widely acknowledged, however, is the notion that the production and exchange possibilities available to the individual, and thus the array of choices he faces, may be further defined by the configuration of the legal and institutional system, the rules of the game. These rules include the property-rights system and the assignment of these rights, as well as the decisionmaking structure and operation of any corporate or collective bodies. The study of these rules of the game has been the traditional domain of the political scientist. He, however, has rarely built his work on a theoretical system resembling that of the economist in analytical power. Only recently has the economist attempted to bring his analytical system to bear on the question of the general economic properties of optimal systems of rules (Cheung, 1969; Demsetz, 1964; Williamson, 1967; Frisch, 1959; Mishan, Pareto Optimality . . . , 1967).

In air pollution economics, little attention has been devoted to these general properties of game rules and the outcomes that they generate. An example is presented in Crocker ("Externalities . . . ," 1971), of a real case where a shift in property-right assignments to the use of the air resource brought about a distinctly different outcome. Varying degrees of attention are given the informational and policing costs of a control agency (see Wright, both works, 1969; Crocker, "Some Economic Aspects . . . ," and "Externalities . . ."). No empirical analysis is attempted, however. In Crocker and Rogers (1971), it is argued that the structure of most air pollution control agencies provides emitters with advantages relative to receptors. However, no formal analysis is undertaken, and nowhere does there exist careful empirical attempts to test hypotheses in this area. In fact, no one seems to have yet published even a thorough description of the political and administrative decision processes involved in the establishment of air pollution control policy.

Customary Discussions

In the absence of serious discussions of the costs and benefits of the fundamental properties of alternative property rights and institutional configurations, the discussion of control instruments has immediately assumed that all property rights in the atmosphere are to be vested in an air pollution control agency of unspecified structure and resources. The control agency is presumed to be omniscient and clairvoyant. Its function is then to employ that control instrument in minimizing the sum of damage and control costs. Under the assumed conditions, economists have opted for the effluent charge (Vickery, 1967; Mills, 1966). It is readily shown, however, even with perfect control agency knowledge, that the charge will oftentimes be impossible or very difficult to calculate. In particular, unless ambient-air concentrations are additive over emitters and unless receptor damages are linear in ambient air concentrations, unique incremental damages cannot be assigned to individual emitters who are but one of a collection of emitters (Davis and Whinston, 1962).

A major fault in discussions of the effluent charge has been their tendency to abstract completely from emitter and receptor location decisions. No discussion of the affluent charge seems to have recognized that the pattern of air pollution damages is not independent of these emitter and receptor decisions (Crocker and Rogers, 1971). However, somewhat more imaginative conceptions of the scope of control instruments have attempted to specify the relations between air pollution dosages, the meterological attributes of the atmosphere, and alternative land use and transportation patterns (Rydell, 1968; Rydell and Stevens, 1968). Without explicit attention to locational decisions, one can reasonably argue that economists can have no idea of what control instrument will minimize the sum of emitter control costs and receptor damage costs.

THE ECONOMICS OF AIR POLLUTION 153

Another factor important to the evaluation of alternative-control instruments is their associated ICP costs. It is these costs, in addition to joint goods or consumption indivisibilities, that cause the air pollution problem. If there were no costs of information or policing, the joint-goods problem would be negated by the formation of a collective control agency with complete power to allocate the air resource. The control agency would be able to obtain instantaneously and without the expenditure of valuable resources the values of all the system's structural parameters. It would also be able to ensure costlessly the compliance of all emitters and receptors. All the agency need do would be to impose on each emitter that emission quantity or--if a facade of voluntarism were thought desirable--that effluent charge which minimized the sum of damage and control costs. Discussions of alternative control instruments without the explicit introduction of imperfect knowledge and the factors which generate imperfect knowledge become rather trivial exercises.

It is the inability of emitters, receptors, and control agencies to secure information and compliance costlessly that forms the nexus of the air pollution control problem. But it is also the case that the manner in which these ICP costs vary with alternative-control instruments has received practically no attention. What, for example, are the conditions under which receptors will resort to the courts rather than attempting to obtain satisfaction from a collective body? For a given property rights configuration, what are the net benefits of increasing the probability of discovering violations of atmospheric property rights relative to increasing the penalties for discovered violations? What are the circumstances, if any, under which the costs of the uncertainties generated by an inaccurate effluent charge outweigh the costs of the rigidities and lags implicit in an invariant, ambient-air standard? How do these ICP costs vary with different property right configurations and assignments? These costs can be expected to vary with a variety of factors: the number of parties who must participate; the information each party needs for optimal decisions; stability of the relevant parameters; stability of participant identities, objective functions and numbers; number of separate transactions required; ease of avoiding contractual arrangements; presence of scale economies in information production and policing; number of distinct input and output characteristics; size of the market for informational and policing services; length of the time interval between planning and initiating an action; length of the time interval over which consequences of an action are brought to bear; the dispersion of possible states of nature; and probably some other factors. Discussions of control instruments must remain irrelevant, for the most part, until intensive economic-theoretic and empirical efforts are made to specify the structure of these ICP costs and the costs due to the uncertainty that they generate.

Assessing Results of Instrument Implementation

The ICP cost problem is a problem in the implementation of control instruments. Examples of actual, successful implementation of anything other than ad hoc instruments are quite rare in air pollution. However, there is an exception: One of the most exciting developments in air pollution economics during the past few years has been the advent of mathematical programing models of the production and cost alternative patterns of regional air quality, paralleling earlier developments in water-resource management (Thomann, 1963; Thomann and Marks, 1966). Fundamentally, these models are complicated empirical applications of the theory of production and cost for the multiproduct firm (Dan, 1966: 148-89). In pollution applications, the firm is an air quality control region (or other duly constituted air pollution control agency), the products are ambient air concentrations of the various pollutants the control of which is the agency's concern, and the plants are known emission sources of these pollutants. An outline of a simple hypothetical model and its elements will perhaps facilitate understanding of the nature and potential uses of these models.

If we use the results of meteorological models that predict the diffusion of pollutant emissions throughout a region (e.g., Turner, "A Diffusion Model . . . ," 1964; Koogler, 1967), the percentage of the total emissions of a specified pollutant from any given source which diffuses to a particular point on a plane defining a geographic area can be obtained.[20]

Let this percentage be given by P_{ijk}^{K}, the percentage of the ith source's emissions of the jth pollutant that diffuse to the kth point. The total contribution of the ith source to ambient air concentrations of the jth pollutant at the kth point is given by

$$P_{ijk} X_{ij} \tag{1}$$

where X_{ij} is the total emissions of the jth pollution by the ith source. Total ambient-air concentrations of the pollutant at the kth point are then given by

$$b_{jk} = \sum_{i=1}^{N_{sj}} P_{ijk} X_{ij} \tag{2}$$

where N_{sj} is the number of sources of the jth pollutant. The pattern

THE ECONOMICS OF AIR POLLUTION

of air quality at a finite number of points of interest may be described by a vector function, the coordinate functions of which are given by

$$\sum_{i=1}^{N_{sj}} P_{ijk} X_{ij} = b_{jk} \qquad \begin{array}{l} j = 1, \ldots, N_p \\ k = 1, \ldots, N_r \end{array} \qquad (2a)$$

where N_p is the number of pollutants under consideration, and N_r is the number of receptor points at which ambient-air concentrations are measured.

The set of equations (2a) completely characterizes air quality for alternative emission patterns

$$X_{ij}, \quad i = 1, \ldots, N_{sj}, \quad j = 1, \ldots, N_p$$

If, in addition, emission-control costs are associated with emissions,

$$c_{ij} = c_{ij}(X_{ij}) \qquad i = 1, \ldots, N_{sj}$$

$$\frac{\partial c_{ij}}{\partial X_{ij}} < 0 \qquad j = 1, \ldots, N_p \qquad (3)$$

where c_{ij} is the cost of controlling emission of the jth pollutant at the ith source, then

$$C = \sum_{j=1}^{N_p} \sum_{i=1}^{N_{sj}} c_{ij} \qquad (3a)$$

where C is the total emissions control outlay in the region. Thus are emission-control outlays completely characterized.

Given emission-control costs associated with the various sources and pollutants, and a desired level of air quality, b* jk, for all pollutants all receptors, the following mathematical programming problem may be formulated:

$$\begin{array}{ll} \min C \\ \text{subject to} & b_{jk} \leqq b^*_{jk} \qquad j = 1, \ldots, N_p \\ & \phantom{b_{jk} \leqq b^*_{jk}} \qquad i = 1, \ldots, N_r \end{array}$$

$$X_{ij} \geq 0 \qquad i = 1, \ldots, N_{sj}$$
$$j = 1, \ldots, N_p$$

To date, several such models have been formulated and solved. In Teller ("Air Pollution Abatement . . . ," 1967), a linear model of the costs of achieving alternative air quality patterns for sulfur oxides using alternative sulfur-content coals is constructed. A multipollutant model that places additional restrictions on the solution values of emissions is available in Burton and Sanjour (1969). These additional restrictions are that each X_{ij} can have only a discrete number of values corresponding to emission levels commonly achieved with currently employed control techniques and that the costs of controlling pollutants at a given source are not independent. A principal concern in both Teller, and Burton and Sanjour, has been the effect on the minimum attained in such a problem of imposing constraints in addition to those dictated by technology and meteorology. Their results indicate that there is a fairly high additional cost attached to the imposition of such constraints (Burton and Sanjour, 1969: 59). For example, the proviso, commonly included in most air pollution control schemes which rely in part on restrictions on the sulfur content of fuels burned within an area, to wit, that all fuel users be subjected to the same restrictions, seems to be particularly costly. Because such additional constraints are frequently embodied in current regional control practices due to equity considerations, Teller, and Burton and Sanjour tend to be critical of current control policy.

In Kohn, ("A Linear Programming . . . ," 1968), is constructed a large, multipollutant model of St. Louis based on simpler meteorological assumptions than those inherent in the work of Teller (1967) and Burton and Sanjour (1969). In general, Kohn's conclusions about the economic efficiency of the practice of air pollution control in the St. Louis area are less critical of current practice than are those of other investigators. The principal reason for this disagreement is the difference between Kohn's meteorological model and that used by the others. The simple premise of Kohn's model is that all sources of a given pollutant contribute in the same proportion to concentrations of that pollutant at all receptor points. Meteorology becomes essentially irrelevant in determining an economically efficient control scheme under this assumption. The necessary condition for least-cost control under this meteorological model is that marginal-control costs at all sources be equal.[21] Kohn found that pollution control officials in St. Louis, by accident or by design, had been quite effective at equating emission-control costs at the margin.

One feature common to all models of this type thus far developed deserves eplicit mention. Note that the air quality production function defined by the system of relations (2a) is linear in emissions. In particular, this excludes consideration of secondary pollutants that are the product of photochemical and physiochemical reactions among primary pollutants. Closely related to this limitation, only stationary-emissions sources have been built into the models with relatively sophisticated meteorological assumptions. However, Kohn's model did incorporate mobile sources.

A PRELUDE TO FUTURE RESEARCH

Historically, even if one concentrates on relatively recent times, measures of the formal participation of economists in air pollution research are not impressive. At the beginning of 1969, of the 17,835 members of the American Economic Association ("Committee on the National Science Foundation Report on . . . ," 1968: 582), it was extremely unlikely that one could find more than 25 members who had devoted any empirical efforts to air pollution problems. Most professional effort appears to be concentrated on relatively traditional economic concerns.

Concomittant with the small input, output also appears to be small. Casual inspection of the extensive bibliography compiled here suggests a singularly small number of publications. It should be noted that the great preponderance of these publications are government reports. Very little specifically concerned with the air pollution problem has appeared in the major professional journals.

Ironically, to the extent that social scientists are supposed to concern themselves, in the main, with pressing social problems, there appears to be a misallocation of professional effort paralleling the misallocation of the air resource. Although there are probably several reasons that can be adduced to explain this apparent scarcity of professional effort, the major reason seems once again to be primarily economic. While it is most assuredly true that the pay today for such work is becoming relatively attractive, there are other considerations embedded in the professional reward system and the costs of doing such work that appear on balance to produce substantial disincentive to work in the area.

There are several substantial costs that accrue to the economist who does research in air pollution and other environmental quality problems. Most obvious is the investment involved in mastering the basic science and technology of air pollution control. To proceed competently, the researcher must have a fair grasp of the material covered in Stern's (1968) three-volume collection of articles on air pollution. In many instances, this background can be attained by formal training available in the engineering schools of most universities or by participation in the training program run by EPA's Air Pollution Office. In general, the researcher should probably plan to spend the equivalent of one full semester of difficult study, and, of course, be prepared to spend time in continuing education.

Thus prepared, the major research cost is the time necessary to bring a project to fruition. The period of production in air pollution econimics appears to be relatively great.[22] A major portion of the human effort is expended in overcoming severe data problems, for there is very little good statistical information bearing on the economic aspects (or any other aspect, for that matter), of air pollution.[23] To make matters worse, so little is understood about the economics of air pollution that the probability of a project failing to measure up to expectations is not negligible. These disincentives are perhaps reinforced to some extent by the academic economist's professional reward system. A high premium attaches to rapid production of scholarly, publishable output.[24]

Nevertheless, there is some evidence that the force of these disincentives has weakened during the last few years. In 1964, a survey indicated that of all the specialties in economics, those in which economists investigating air pollution were likely to fall (land economics and welfare), were among the most poorly paid specialties (Committee on the National Science Foundation . . . , 1968). More recent data are not yet available, but there is some reason to believe that these salary rankings have been reversed. However, a casual inspection of available data on research expenditures in the economics of air pollution makes it seem likely that the increase in demand for economists' services evidenced by salary increases is concentrated in environmental quality problems other than air pollution, e.g., water pollution. Some would assert that the relative lag in air pollution economics is due to the stranglehold people with backgrounds in medicine and engineering have had on air pollution research policy. In any case, the research funds allocated to air pollution economics are piddling in comparison with the seeming enormity of the air pollution problem.

Through 1970, NAPCA's current annual budget was about $100 million.[25] Of the total budget, $31.9 million was being allocated to research grants and contracts. About $2.4 million of this research budget was devoted to economic investigations, a perhaps not unreasonable proportion. Five eighths of this $2.4 million ($1.5 million), was devoted to emissions-systems studies, a good many of which appear to be the previously mentioned piecemeal engineering studies to which economic analysis is more or less incidental. One lonely $102,000 study by a political scientist was devoting attention to legal and institutional constraints, and three or four other studies consuming no more than $500,000 were investigating regional air quality management systems. Perhaps the best indicator of the status of air pollution economics research is to be found in an NAPCA publication ("Guide to Research," 1970), where, among nineteen categories, economics studies were placed in a category entitled "Miscellaneous Studies-- Economics, Statistics, and Legal."

It is apparent that relatively little in the way of resources is today being put into air pollution economics research programs. This brings us back to the difficult question: "Do we really care?" This question, like any other similarly difficult question, can only be answered legitimately with the somewhat irresolute: "It all depends."

First, it depends on the potential payoff involved in effecting air quality improvement using findings developed as a part of the research program as opposed to effecting such imporvements unaided by research results. To our knowledge, there is little available evidence on which to base an evaluation of the payoff. However, in certain isolated instances, it has been demonstrated that economic research can indicate new ways of approaching abatement which are far superior insofar as purely pecuniary considerations are concerned, e.g., Ernst and Ernst ("Cost-Effectiveness Study," 1969). This one research finding, if implemented in even one city, pays many times the research cost in decreases in emission-control outlay needed to achieve a given air-quality standard.

Another instance that could be cited is the potential usefulness of the results of macroeconomic studies. Never before has the waste-disposal problem been looked at in total as suggested in Ayres and Kneese (1969). The results of such a total look may produce startlingly different control policies. The results of forcasting-model studies (Anderson, "A Note on . . . ," 1970b), may be used to combine appropriate economic policies to offset, insofar as possible, any undesirable economic consequences of control.

Second, the payoffs from a research program depend on the extent to which research results, regardless of how potentially valuable they may be, are or are not used. Today, air pollution control tends to be the province of engineers and physicians. In addition to accumulation of vast stores of valuable knowledge over their decades of experience, some strong traditions and prejudices have also developed. Because, as was observed in Boulding (1967), it is the role of the economist to question tradition, professional cooperation is strained by differences in approach. This is especially true where the engineer can justifiably assume, as is frequently the case, that the economist is abysmally ignorant of the basic science of air pollution and its control. If cooperation is not much improved over what it appears to be today, there is little hope that even the soundest of findings will significantly influence policy. Much attention needs to be given this particular vital link in making a case for the worth of an air pollution economics research program.

Although we do not pretend to be completely disinterested, we believe that a balanced consideration of the questions raised above suggests that the practical consequences of an applied economics research program of the kind currently envisioned is likely to yield substantial benefits to the people of the nation. Careful direction of research to assure that high potential payoff areas receive priority, coupled with an effort to develop close effective ties between economic research and air pollution control personnel, should suffice to make this a safe prediction. The economists' experience in studying highly interdependent and largely nonexperimental phenomena should prepare him to be a valuable member of an applied-research effort. By training, he is prepared to work not only on economic problems, but on more general problems where mathematical modelling or statistical measurements are involved. We believe that he should become so involved.

With regard to research priorities, we would stress the development of more knowledge about the emitters and receptors and more knowledge about the efficacy of alternative-control instruments. We would not stress the development of so-called systems models. It is our conviction (in the absence of fortunate but unpredictable countervailing errors), that systems models are only as good, in a practical context, as the component parts on which they are based. Our reading of the literature reporting what is known about these component parts suggests that really very little is known about them.

The reasons for stressing research on control instruments are even more apparent. It seems to us that air pollution is primarily a social problem, not a technical one. One of the favorite concluding lines of afterdinner speakers on air pollution topics is that we have today, as we have had for many years, the technical means to control air pollution satisfactorily. What we lack are effective institutional means of weighing the technical alternatives. The issues raised in the introductory section and in the control-instruments section on ICP costs and property rights are clearly crucial to effective, long-term eradication of air pollution problems.

NOTES

1. An overview that conveys the apparent cyclical nature of these movements is to be found in Murphy (1967). Their intellectual

antecedents are carefully set forth in Barnett and Morse (1963: 17-95). This latter book is absolutely fundamental to any serious study of natural resource economics.

2. At present, there does not exist any rigorous economic-theoretic development of these notions. Mishan (1967), in an admittedly polemical piece, develops them at some length. Boulding (1969), employs a successful imagery. An earlier version is available in Scitovsky (1959). Presentations intended to serve only as a justification for the theoretical development of a general equilibrium approach to environmental management introduce Kneese and d'Arge as well as Ayres (1969).

3. The relevant references are highly technical. Among the more important are McKenzie (1959), Arrow and Debreu (1954), and Debreu (1959).

4. A detailed derivation of the necessary conditions reveals, as an intermediate step, a system of homogeneous eqautions having the property of too many solutions (infinitely many), or too few (none).

5. An example where the market does not fail to allocate the air resource is given in Dolbear (1967). However, the disincentive to coalition formation occasioned by the near impossibility of excluding one's neighbors fails to arise because the example employs only one emitter and one receptor.

6. Informational cost means the cost of obtaining the information about the attributes of the goods in question and the state of nature necessary to enter into bargaining or market transactions. Contractual cost involves the cost of finding the market or someone with whom to bargain as well as the costs associated with the actual transaction. Included would be the costs of forming and maintaining coalitions with fellow buyers or sellers. Policing costs are simply the costs of ensuring that the terms of a transaction, once made, are adhered to.

7. The distinction between bargaining and market transactions is quite intentional, although the difference is a matter of degree rather than kind. In bargaining, contract attributes other than price are presumed to play an important part. For example, a contract arrived at by bargaining between individual receptors and emitters will likely include, in addition to price, stipulations about the contract's time interval, the types of inputs each party may employ, procedures in case of default, distribution of policing costs, and assorted other matters. In the case of market transactions, the major element in exchange is price. Other attributes of the exchange, e.g., policing, are provided by some collective body and are not subject to negotiation

during the exchange process. Of course, the presence or absence of these collectively supplied attributes will be reflected in the exchange price.

8. Formally, a lexicographic ordering is defined as follows. Let
$$W^1 = (W_1^1, \ldots, W_n^1) \text{ and } W^2 = (w_1^2, w_2^2, \ldots, w_n^2)$$
be any two n-dimensional vectors. W^1 is defined to be lexicographically less than W^2 if and only if
$$w_m^1 \quad w_m^2$$
for the smallest integer m $(0 < m \leqq n)$
for which $w_m^1 \neq w_m^2$. Such an ordering cannot be represented by any numerical function, whether continuous or not, in the services the air resources provide (Pearse, 1964: 23-24).

9. For example, the National Air Pollution Control Administration ("Guidelines," 1969: 12) states: "Particular consideration must be given to the public-health implications of any air quality standards proposed for adoption. Air-quality standards that do not reflect such consideration cannot be approved." A glance at the ambient air standards being adopted for sulfur oxides and suspended particulates in the federal air quality control regions confirms the assertion that prevention of health effects to even the most susceptible parts of the population consistently receives the highest priority.

10. For example, two emitters may have different output targets that may or may not be realized. The control equipment they install, and thus, its operating cost, will vary according to the targeted output as well as the realized output.

11. For a succinct presentation of the differences between the engineering and the classical economic view of the production function, see Smith (1966: 200-18). The current economic emphasis on activity analysis to analyze production relationships would seem to have reduced this disparity somewhat. Both approaches now employ design parameters as the choice-theoretic unit.

12. For an argument that the welfare effects of control options are not independent of market structures, see Buchanan (1969).

13. The market for coal that has had its pollutant constitutents partially removed is considered in Jimeson (1965). Some recognition

THE ECONOMICS OF AIR POLLUTION   163

is taken in Crocker ("Some Economic Aspects . . . ," 1968), of the
influence of the structure of the phosphate industry on its firms'
pollution-control policies.

14. It is worth noting that emitters and their representatives
often seem to take as a basic but implicit premise that any control
policy which alters the competitive structure of the industry is undesirable by its very nature. For an example, see Weber (1970).

15. In many circumstances, the individual can reduce the uncertainty he faces by participating in contracts having payoffs contingent on the occurence of possible world states. However, there exists
a wide variety of contingencies in air pollution situations, and it is
probably impossible to make contingent contracts for all of them.
To the best of the writers' knowledge, there are no real world examples
of insurance or contingent-claims markets in air pollution events.

16. Agricultural economists have sometimes noted the discrepancies between experimental and real yields of commercial crops.
See Swanson (1957), for example.

17. A good starting point for an investigation of this kind is
Schelling (1968), and Zeckhauser (1969).

18. In Anderson and Crocker (1969), the point is made that it is
totally irrelevant whether or not the receptor perceives differences
in air pollution dosages. All that matters for damage measurement
is that he perceive the _effects_ of these dosages. The notion of cause
and effect need reside only in the mind of the investigator. This
statement is correct if and only if receptor perception of the link
between dosages and effects is neutral with respect to the adjustment
possibilities receptors perceive and select.

19. This is a bit too strong. Distributional questions and therefore some questions of allocative efficiency would remain. When one
does introduce distributional considerations, a rather large number
of as yet uninvestigated research questions become apparent. For
example, how does the quantity and the value of waste production from
consumption activities vary over income classes? When a collective
body is made responsible for air pollution control, what is the incidence
of the taxes necessary for financing the body's activities?

20. Since ambient air concentrations are usually expressed on
a weight/volume or volume/volume basis, the P_{ijk} are not pure
dimensionless percentages. They depend on the dimensions in which
air quality is measured.

21. Let $P_{ijk} = \varphi_j$ for all i and k.

If the constrained minimization problem were approximated by a classical calculus constrained minimization problem as

$$\min \sum c_{ij}(x_{ij}) + \sum \lambda k \sum (\varphi_j x_{ij} - b^*_{jk})$$

$$\frac{\partial c_{ij}}{\partial x_{ij}} = \sum \lambda_k \varphi_j, \text{ for all i.}$$

The first-order conditions evidently require that the marginal cost of control of all sources of the jth pollutant be equal.

22. The examples with which are most familiar (Anderson, 1969; Anderson and Crocker, 1970; Crocker, "Some Economic Aspects ...," 1968), as best as we can recall, involved the following resource expenditures over and above the time spent learning fundamentals.

	Anderson (1969)	Anderson and Crocker (1970)	Crocker (1968)
Principal Investigator	1 man/year	3/4 man/year	1-1/4 man/year
Assistance	1/6 man/year	1/3 man/year	1/2 man/year
Computer time			
IBM 360-65	3 hours		
CDC 6500	1 hour		
CDC 3600	1 hour		
UNIVAC 1108	1/2 hour	1½ hours	1 hour

23. Of the expenditures cited in the note 22, almost all the assistance and about one third to one half of the principal investigators' time was spent on data problems.

24. These disincentives seem to be reflected in the budget of Resources for the Future, Incorporated, a private research organization devoted to the economic and social aspects of natural-resource problems. In their annual reports (1967, 1968, 1969), a good deal less than 10 percent of the organization's grants in 1967 and 1968 could be identified with air pollution projects. In 1969, this allocation had increased to over 20 percent.

25. Larry Barrett, formerly of the Office of Program Development, NAPCA, and Brian Peckham, formerly of the Division of Economic Effects Research, NAPCA, generously provided us with this budgetary information. However, the opinions expressed are solely the responsibilities of the authors of this chapter.

CHAPTER

7

**CONCLUSION:
A POLICY TRADE-OFF
ANALYSIS MODEL
FOR AN AIRSHED**
Paul B. Downing

INTRODUCTION

This final chapter summarizes the preceding material in a way which places emphasis on its policy implications. It uses this material and information from other sources to develop a framework for the analysis of alternative policies available to decisionmakers. Hence, it describes a policy trade-off analysis model. It attempts, in a very preliminary way, to interrelate the physical, economic, social, administrative, and institutional aspects of problems that affect and are affected by any control policy.

The full implementation of a model such as the one suggested here is at least several years away. Many of the subsections of this model have not been fully conceptualized, let alone developed. This effort can be thought of as a suggested direction for the development of such a model. But this discussion has a more immediate use. It suggests the many aspects of the air quality control problem that ought to be considered by decisionmakers. It indicates those areas that the author considers to be most important for policy analysis and suggests their interrelationships with one another.

The quantification of such a model would enable analysists to present decisionmakers with a rough approximation of the full implications of each alternative policy under consideration. The end product of such an effort would be an analysis of several policy alternatives in a way that would make clear the trade-offs between competing social goals of each alternative. The decisionmakers would than have before them, for the first time, all the technical and socioeconomic information they should consider in order to be most effective in their policy deliberations. They would still have to make the difficult decisions about how much to weight each positive and negative aspect of the alternatives in order to select that alternative which they deemed best for the people of the airshed. The problems of implementing such a model will be discussed at the end of this chapter. They are sufficient to suggest that a full quantification of the model in the simulation sense is not likely in the near future. Rather, the author envisions a set of submodels that are loosely (perhaps verbally), connected. In other words, several partial equilibrium models would be developed, and analysists would judgmentally quantify the interrelationships among them for any policy alternative. The important thing is the inclusion of all the major aspects of the alternatives and their affects on decisions. It is obvious that no one discipline can accomplish the above task. It should be equally obvious that physical and biological scientists are not likely to produce a meaningful model of the airshed without a very substantial input from the social sciences.

Implicitly assumed in this approach is the possibility of a governmental administrative agency with sufficient jurisdiction to cover all parts of an airshed and with the power to act. (Chapter 4 suggests that this situation does not now exist and is likely to be difficult to attain). We also assume that this administrative agency is responsive to the wishes and desires of all the people in the airshed rather than reflecting heavily the point of view of the polluters as has been the case in many existing agencies. (The Nader Study Group seems to take this position; see Esposito, 1970.) We also assume a legislative body that is itself responsive to the wishes and desires of all the people. In other words, we assume a system that fairly adjudicates between the wishes and desires of different groups competing for use of our air resource. (See Hagevik, 1970, Chapter 7, for a discussion of the problem of control strategy and the role of bargaining.) To the extent that the above assumptions are not valid, the following model is less useful. However, we believe, perhaps naively, that public access to information on air pollution control alternatives will lead to a better (more rational) and more effective (publicly accepted and implemented) control effort. Facts make incorrect or biased decisions obvious and arouse public pressure for change.

BASIC ISSUES AND SUGGESTED DECISIONMAKING METHODOLOGY

There are two basic issues in air-quality management about which society must decide. One issue is the degree to which society wishes to bring about the reallocation of resources necessary to effect an increase in control. A correlary to this issue is the decisions

CONCLUSION 169

concerned with where resources will be withdrawn from alternative uses and who will pay for the reallocations. It should be clear that all of these issues are interdependent. A suggested methodology for the analysis of policy will be presented in this section. Discussed in greater detail below, are the following: the policy trade-off analysis model, the interrelationships among sectors of the model and hence among the elements of the decisionmaking process, and the desired level of quantification of the model.

The Level of Air Quality

Let us turn to a discussion of the basic issue first mentioned. In several chapters of this book, it has been argued that clean air (air with only natural or background levels of the key compounds), may be the ideal, although society probably cannot reach this level technologically at present. It has also been argued that even if we could reach this air quality, the resources used and disruption of society would not justify such an expenditure. These chapters have also pointed out that the currently prevailing levels of air quality are in many cases likely to be poorer than is desirable from the point of view of society. The question then is: What level of air quality between the present ambient-air quality and natural-background levels should be obtained in any one airshed?

Before discussing one basis on which this question can be answered, we want to make an additional point about air quality. The quality of the air is not one measure but measures of several compounds. Some of these compounds are emitted directly from stationary or mobile sources. These are called primary pollutants. The five principle, primary pollutants are carbon monoxide (CO), hydrocarbons (HC), nitrogen oxides (NO_x), sulfur dioxide (SO_2), and particulates. One of the particulates that has recently been receiving special attention is lead.

In addition, these primary pollutants react with one another and with other compounds found in the air to produce additional compounds. These are called secondary pollutants. The one example most often cited is the reaction of HC and NO_x in the presence of sunlight to form ozone (O_3). This photochemical reaction produces the smog that Los Angeles has made famous. However, this reaction is cause for increasing concern throught this country and in other countries as well. Of growing concern among scientists are some of the other secondary reaction products of the photochemical process. These include peroxyacyl nitrates (PAN), and peroxybenzoyl nitrate (PB_zN) both of which have been shown to be strong eye irritants (Stephens, 1970).

From the discussion of the above two paragraphs, it is obvious that air quality is not one good but several goods. Each city has a different composition of air quality, and this composition varies among the sections of the city and during different times of the day and year for any one section. Thus, we will refer in the remainder of this chapter to air-quality sets that are combinations of the levels of air-quality parameters such as NO_x, variations in their geographic distribution, and their variations over time. Because much of the damage from air pollution is a function of the length and duration of peak levels, the air-quality sets are assumed to stress this aspect of quality.

The criterion we suggest for deciding on an air-quality set to be reached in an airshed is a variant of the traditional benefit/cost analysis methodology used in water-resource decisionmaking. The traditional benefit/cost theory states that one expands a government project (in this case, increased air-quality control), until the additional benefit of one more unit of output (control), is just equal to the additional costs incurred and that any project whose total benefits exceed its total costs be undertaken.

The employment of this criterion in the above form depends on our ability to completely and accurately quantify in money terms all the benefits and costs of each possible project. However, by their very nature, most public activities have benefits and sometimes costs that are very difficult and/or costly to determine. This leads us to suggest the following procedure: First, we suggest that a variety of air-quality sets be determined. For each of these sets, the benefits in terms of damages avoided are to be quantified to the degree justified by the cost and benefits of quantification. For those items where it is possible and desirable, the quantification should be in money terms. For those cases where benefits cannot be quantified in money terms, some physical quantification should be attempted. Failing this, the benefits should be quantified in subjective terms. In this way, all benefits will be quantified as completely as is practical. While it is true that costs of control are more easily quantified than benefits, some costs are not amenable to quantification in monetary terms. Thus, some physical and subjective quantification of costs will also be necessary.

Having quantified both the benefits and costs of alternative air-quality sets as completely as practical, the results, which comprise a group of policy alternatives, are presented to decisionmakers. The decisionmakers will employ their subjective judgements of the public's desire for cleaner air relative to other social goals to select that air-quality set which appears to maximize social welfare.

CONCLUSION

How to Obtain Desired Air Quality

There are three principle parts to the question of how to obtain the desired air quality. The first is: How can society achieve the desired air quality in a technically feasible and economically efficient manner? The issue is essentially one of quantifying the costs of alternative controls and selecting those controls that are economically efficient at reaching specified air-quality goals. To do this, a cost-effectiveness study is proposed.

The second part of this question is: How can society obtain this level of control administratively and with what institutions and instruments? Decisions here focus on different control instruments, problems of enforcement of controls, and the cost of enforcement and administration.

The third part of this question is: Who should pay for improvements in air quality? This is a social value judgment. We will, however, have some things to say about the effect of other decisions on who pays and the effect of attempts to change the financial burden on the control effort.

Interdependency of Decisions

The two basic issues discussed here are neither independent of one another, nor does one precede the other in time. Rather, all these issues are the main parts of a simultaneous system. Decisions in one part of the system effect other parts in one way or another.

In the remaining sections of this chapter, we will present a model of the interrelationships among different factors that affect the outcome of any air pollution control policy. The model is an attempt to identify the key elements of the system and their relation to the goal of air quality. The model focuses on the modified benefit/cost analysis methodology suggested in this section.

LEAST COST, AIR-QUALITY SETS

The goal of this section of the model is to develop various technically feasible and economically efficient air-quality sets. These may be termed the least-cost air-quality sets in the sense that a given air-quality set is reached at the lowest possible money cost. In conjunction with these sets, the related secondary effects of each alternative would be determined. These secondary effects include the cost of effects on other environmental areas such as water pollution, the danger of a nuclear power plant accident, effects of control efforts on income distribution, effects of control costs on firm output and employment, and increases in demand for other goods and services such as electricity.

Six elements comprise the inputs to a least-cost emission-control model (Figure 1). The air-chemistry and meteorology-simulation model relates primary emissions to observed air quality. Factors that affect the ambient-air quality at any point in time and space include the kind of pollutants released, their amount, the location of their release, the timing of their release, and the meteorological conditions. This would seem to be a straight-forward problem, but many complications arise. First, chemical reactions that take place in the atmosphere are very complex, and the secondary pollutants produced are unstable, so that the entire process is only partly understood. Second, meteorological conditions and air movements are also complex and change rapidly. The combination of these two complexities yields an only partially understood system that is different for each airshed. The output of this model is an estimate of air-quality parameters by location and timing, given a specified set of meteorological conditions and a specified set of emission sources, amounts, characteristics, and timing of release. With this model, we would be able to estimate the improvements in air quality resulting from various control possibilities.

There are various technical-control alternatives that might be employed to control one source. Each has some ability to control the release of one or more primary pollutants at some cost. These technical controls can include add-on devices such as stack precipitators for power plants and catalytic reactors for cars. They can also include major process shifts and plant reconstruction or shutdown. Some of these controls may reduce the release of some pollutants while increasing the emission of others. The costs associated with these control alternatives are included in the "direct costs of control alternatives."

In addition to technical-control alternatives, there are control techniques that directly affect the output, location, and timing of primary pollutants. These controls we term "social-control alternatives." They include such things as emergency measures that temporarily restrict or shut down polluting plants and restrict or stop automobile movement during periods of particularly severe air pollution. They can also include milder forms of controls such as changes in the work schedules of plants and offices in order to change the timing of pollutants' release. Land-use controls would also be included. To

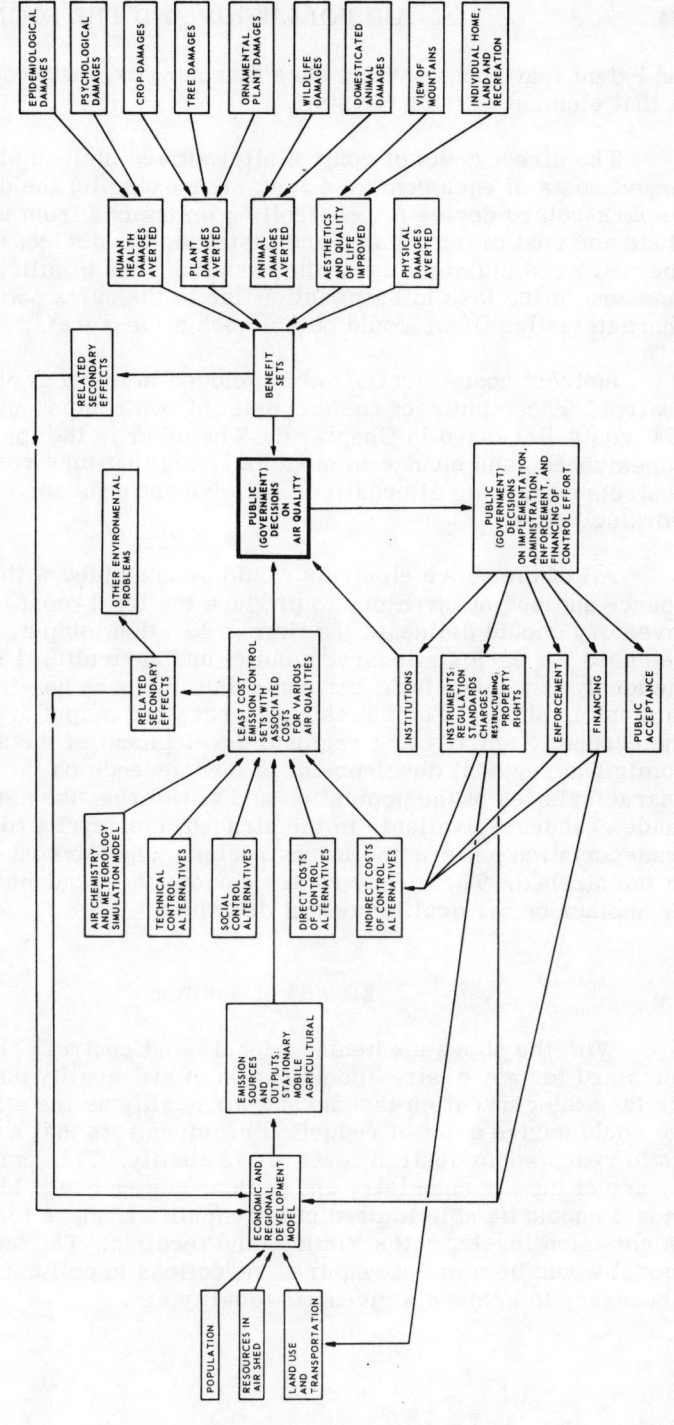

FIGURE 1

Air Pollution Control Policy Trade-off Analysis Flow Chart

the extent that such controls have direct costs, they would be included in that element of the analysis.

The direct costs of control alternatives include all the out-of-pocket costs of each control effort. For example, the direct costs of a spark-retard device for controlling emissions from used cars include the cost of purchasing and installing the device, the cost of operating and maintaining it, the resultant loss in milage, and some measure of the loss in satisfaction due to the car's poorer performance characteristics (if we could obtain such a measure).

Indirect costs must also be included in the cost of air pollution control. These indirect costs consist of two components. One is the ICP costs discussed in Chapter 6. The other is the loss of social cohesiveness and change in mode of living that may result from such controls as forcing alternative land-use patterns and changes in working hours.

All of the above elements would be combined with an emission source and output inventory to produce the least-cost estimates. The inventory should delineate the timing, location, output, and pollutant released for each stationary, mobile, and agricultural source. This obviously is a large task, but some data for such an effort is available in most airsheds. The emission sources and output are dependent on the economic activity and regional development of the airshed. Economic and regional development in turn depends on the level and characteristics of the population and workforce, the natural and man-made resources available in the airshed, the developed land use, the transportation pattern and infrastructure, and the cost of doing business in the airshed. These factors are obviously dependent to one degree or another on air quality control decisions.

Method of Solution

With the above elements, a least-cost control set could be determined for any desired combination of air-quality parameters. If we take the current emissions and air quality as the starting point, we could derive a set of reduction requirements that would enable us to reach some desired level of air quality. This could be done by use of the air chemistry and meteorology model. Ideally, this model should be able to predict air-quality changes for various changes in emission levels, rates, timing, and location. The output of this model would be a set of required reductions in pollutant emissions necessary to achieve a given air-quality set.

CONCLUSION

From the technical and social control alternatives, we would derive a listing of devices and other controls applicable to one or more sources. Each such device or control would reduce the emission of one or more pollutants (and might increase the emissions of another). Each device or control would also have an estimate of its associated cost that included both direct and indirect costs.

The object is to find that combination of devices and other controls applied to various sources which will reach or exceed the required reductions at the least cost. This can be done by using a linear programming model. For one use of linear programming to estimate least-cost controls for meeting an air-quality standard, see Kohn (1970).

Different sets of reductions could be fed into the linear programming model and the resultant costs and device utilization determined. This is what is meant by the "least cost emission control sets with associated costs for various air qualities," referred to in Figure 1. However, it must be kept in mind that not all costs will be capable of being quantified in a way which allows their inclusion in this model. Such costs would be quantified in some other way and included in the associated costs.

Related Secondary Effects

As has been suggested before, air pollution control efforts are likely to have certain side effects. One main group of these secondary effects is the effect of air pollution control on the release of pollutants to other environmental media such as water or land. These effects should be considered lest we find we have not solved our problem but merely shifted it to another form. The trade-off between air quality and the quality of other elements of the enviornment is one that can and should be made explicit. It may in fact be rational to say that we should increase water pollution in a location in order to gain cleaner air. This would be so if the sum of air pollution control costs incurred and damages averted by the cleaner air were greater than the sum of water pollution control costs incurred and damages caused by the dirtier water (and society did not want to buy both clean air and clean water).

The second group of secondary effects relate to the effect of air pollution control costs on the economic development of the region and hence the output of pollutants by firms and individuals. An increase in the cost of production due to air pollution control requirements may cause a firm to reduce output and employment. This is likely to reduce emissions from that source and perhaps reduce emissions from automobiles because fewer people were driving to work at the plant. At the same time, the benefits of improved air quality would make people more willing to live in the airshed. This may reduce wage demands and increase employment. The net effect of these two forces and their multiplier effects is dependent on the individual circumstances of each airshed and each polluter.

BENEFIT SETS

After we have determined the cost of obtaining various levels of air quality, we would then wish to determine the associated benefits of having attained that level of air quality. For our purposes, we have defined benefits in a negative way as damages averted. A damage is any undesirable result of a given air quality, e.g., the damages averted when a reduction in the average oxidant reading over a five-year period reduces the percentage probability of an individual contracting emphysema. These damages could be quantified in terms of treatment costs, work-time lost, and so forth (see Ridker, 1967). We have divided the benefit set into five main categories: human health, plant damages, animal damages, physical damages, and aesthetics and quality of life. Again we would start from the existing air quality in the airshed and treat any reduction in damages as a benefit to be related to the cost of obtaining that improvement. Note that some air quality parameter readings may increase and cause more damages. These would be treated as negative benefits rather than costs. It should be noted that this measure of benefits is a lower bound in that it does not measure the "willingness to pay" to avoid the damages (Schelling, 1968).

The first and probably most important category of benefits is human health damages averted. Health problems are often the overriding concern in efforts to control air pollution. Health effects can be divided into two general areas of concern: epidemiological effects and psychological effects. The epidemiological effects of air pollutants are the disease-causing effects. One such effect is the increase in cases of emphysema with increases in air pollution. This disease is apparently caused by the combined presence of SO_2 and particulates and may also be caused by oxidant or NO_2. Carbon monoxide has been shown to reduce the oxygen-carrying capacity of the blood and to cause increased incidence of heart disease. Many such effects have been shown. These effects vary with the level and duration of exposure to various pollutants. It is also true that the health effects of exposure to two pollutants at the same time may be much greater than for each one alone (Timothy Crocker, 1970).

CONCLUSION

The possibility of phychological damages from air pollutants was discussed in depth in Chapter 3. This relatively neglected aspect of human-health damages may justify even lower levels of air pollutants than the epidemiological effects discussed above. This is true because psychological damages such as reduced-reaction time appear to occur at much lower levels of pollutants than epidemiological damages.

Damages to plants due to air pollutants have been widely demonstrated. Pollutants can cause a reduction in growth and yield of crops and trees. They can also cause visible damage to crops, trees, and ornamentals that reduces their salability and the aesthetic pleasure one derives from viewing them. Some levels of pollutants have been reached that have resulted in the death of plants. Animals are also subject to reductions in growth, salability, and life span due to air pollution; however, except in a few specific cases, these effects are far less documented than are plant damages.

There are many physical damages attributable to air pollution. One such damage is the increased soiling of objects such as clothing and buildings. This causes a greater desired frequency of cleaning and perhaps a reduction in useful life. Pollutants can also cause deterioration of compounds such as nylon and rubber. The extent of these physical damages has only been roughly approximated, and much work remains to be done. Each primary or secondary pollutant has its own peculiar physical effects.

The last category of benefits is aesthetics and the quality of life. Perhaps the most noticable aesthetic effect of air pollution is reduced visibility and general haze. This is often caused by aerosols produced in the photochemical reaction. This haze has a general, depressing effect on people. Pleasurable views of distant objects such as mountains are lost. Other aesthetic problems that come to mind quickly are the smells associated with high ozone levels, high SO_2 levels, and other specific sources such as paper mills and tanneries. Perhaps it is these easily perceived damages that are most responsible for the public outcry to clean up the air. One example of how the public outcry may be misdirected for lack of the facts is a steel plant in the Los Angeles airshed. This plant produces coke to use in iron smelting. When the coke is released from the ovens, it is quenched with water. This results in great clouds of water vapor rising from the plant periodically. People cite these clouds as air pollution, and one television documentary used it as an example of the pollution still being released in the airshed. In fact, the visible cloud is harmless water vapor. The harmful emissions of this plant, which are numerous, are not visible. One wonders if the public would be satisfied if these visible releases were eliminated, but the invisible damaging pollutant release were not. If the damages are primarily psychological, perhaps this is what should be done.

In addition to aesthetics, the quality of life is affected in other ways. These center around the individual's home or recreational activities. People find that vegetation is not as lush and green as it might be without air pollution. They also find, during some times of the year, that they must stop their childern from playing too hard because high consumption of air is not good for them. At such times, adult recreation is also curtailed. People may drive or move somewhere else to get away from the effects of air pollution only to find that it follows them. It is difficult to separate aesthetics from the other four categories, but it is the most commonly cited reason for cleaning up the air.

As with the cost of our model, the various benefit sets resulting from the air-quality improvements being analyzed produce some secondary effects on the system. Improved health, aesthetics, and so forth are likely to increase the tendency of people to locate or remain in the airshed. Their presence will increase the demand for goods and services, the production of which causes air pollution. Also, with cleaner air, options and activities are less constrained so that some change in the economic structure of the area may occur. For example, improved air quality may increase the demand for recreational services, and a relative shift in industry mix and land use will occur. Such effects could be many and varied so that anticipation of most of them may be nearly impossible. These effects may partially or wholly offset the bad secondary effects felt on the cost side.

ADMINISTRATIVE AND INSTITUTIONAL STRUCTURE OF CONTROL EFFORTS

One might conceive of the decision process as consisting of two parts. The decisionmakers have decided upon an air-quality set on the basis of the information produced within the framework of the sections above. In doing so, they have selected automatically the most economically efficient set of controls for that quality. All that would remain to be determined would be the method of implementing their decisions. But things are not quite that simple, for the decision on air quality can depend on who the decisionmaker is, and this can depend on the administrative structure that is in force at the time of the decision.

At the begining of this chapter, we set out an assumption about the decisionmakers. We assumed that they had jurisdiction over the entire airshed, that they had the power to act, and that they fairly

CONCLUSION

adjudicated among differing points of view held by the population affected. The design of an institutional structure that would even roughly bring about such a decisionmaking body is a very difficult task.

First, we already have a multitude of institutions in existence that have overlapping jurisdictions both in terms of area covered and type of pollutant or source controlled (see Chapter 4). It would appear that this sort of system, with its conflicts and confusion in jurisdiction, has not worked well for the people affected.

Second, polluters are generally more powerful (both economically and politically), than are recipients. This is true for several reasons explored previously. To design a system that counterbalances their "natural" power with an equal amount of recipient power would be very difficult. Furthermore, its adoption faces the same difficulty because it must be approved by those who currently hold the decisionmaking power. We do not suggest that current institutions are completely unresponsive to the effects of pollution. We merely argue that the relative weighting is biased against control. Costs are weighed more heavily than benefits. (There is a similar although less intense bias in our approach, in that costs are explicit, immediately felt, and relatively, easily quantified, while benefits are more tenuous, less immediate, more diffused among people, and difficult to quantify.)

While the design of unbiased institutions for air-pollution control is difficult, jurisdictional problems are more easily dealt with, at least conceptually. If jurisdictions are smaller in area covered than the entire airshed, a less than efficient control effort is likely. Each institution has the incentive to blame the others for not solving the problem. It leads to such statements as: "Your air pollution problem does not come from the area over which I have power." Each administrator tries to make himself look good while, at the same time, he tries not to upset too many people in his jurisdiction. The same sort of game occurs when jurisdictions are split among sources. Each administrator is likely to argue that he has his part of the problem solved, and if only the other guy would do his job, air quality would start improving. These arguments are obviously fruitless and serve only to divert public and administrative attention away from the true issue of control.

The obove argument leads us to recommend an air pollution control institution that covers the entire airshed and all sources within it. But this does not solve all of our jurisdictional problems. The federal government as well as state and local governments have taken some responsibility for air pollution control. This leads to questions such as who has jurisdictional control over a regulated

emitter and which standard is to be followed. The trend seem to be
a national standard for industries that in some cases can be made
more stringent by a local agency, but in no case can they be made
less stringent. The argument for this form of regulation is that with
it no firm in an industry will be at a competitive disadvantage because
it has to control its emission to a greater degree than other firms,
nor will there be the incentive to move a plant from one location to
another simply to escape air pollution controls. However, a case can
be made for doing exactly what this arrangement seeks to avoid. If
the people in one airshed were willing to put up with a decrease in
air quality in exchange for the employment generated by the firm,
they should have the right to do so. This can be justified both in terms
of political rights to self-determination and in terms of economic
efficiency.

Once we have decided on the institutional form of the agency, we
would have to decide on the instruments to be employed. They can be
divided into regulations, standards, charges, property-rights restructuring, and others. Because these instruments are outlined in Chapter
1 and discussed in detail in both Chapter 5 and Chapter 6, they will
not be discussed here. Suffice it to say that the choice of instruments
is neither independent of the problem of enforcement, nor is it independent of the effectiveness and cost of alternative-control devices.
The choice also has implications for the financing of the control effort.

Enforcement is an often neglected aspect of the control effort.
One part of the enforcement problem is the difficulty of obtaining
measures of the actual emissions form various sources. In many
cases, instrumentation is not now available that would quickly and
accurately measure emissions. It is particularly difficult to obtain
reliable and roughly accurate instrumentation that records on a
continuous basis. For stationary sources, one of the major problems
is obtaining access to the source without letting the plant management
know it is being monitored. In too many cases to be taken as chance,
control equipment is "broken down" on the day of the visit. For mobile
sources such as automobiles, periodic, fairly rapid checks are possible
but are rarely employed (an exception is the annual inspection system
employed by the State of New Jersey). This is true because of the
supposed unwillingness of people to submit to inspection, a fact refuted
by survey work done in California (Gold, 1970), and lack of authority
and/or funds for such a program. Yet it is known that 60 percent
of the automobiles in California exceed the standards for their year
(d'Arge, 1970). A simple inspection procedure would not be expensive
and could substantially reduce emissions from this source (Downing
and Stoddard, 1970). The legal instruments used should take enforcement problems into account.

CONCLUSION 181

A second major problem with enforcement is that it is expensive. The inspector must gather information on violations by frequent site visits. Once a case is built, it is a long and expensive legal process to convict a firm of a violation. The legal structure now employed causes great difficulty for enforcement (see Chapter 5), and even if the firm is convicted, penalties have often been insignificant.

There are two issues involved in the subject of financing air pollution control activities. One is the adequacy of the control agency's budget for administration and enforcement. It is not uncommon to find an agency with insufficient funds to adequately administer and enforce regulations. Some efforts can be made to more efficiently use those funds now available by designing probability-sampling systems for enforcement (Chapter 6). But the administrator is constrained from doing too good a job of enforcement lest he alienate the legislative body from which he receives his money if that body is responsive to pressures from the affected polluters.

The second issue in financing is who pays (provides resources), for the control effort. There appear to be three choices: the polluter, the recipient, and the general public (through government). The argument in favor of charging the polluter is that he causes the pollution with his action and should be required to pay for its control: It is his fault. In the case of polluting firms, it is typical that the firm will not pay the full cost of the controls or process changes it is required to install. Instead, it will pass part of this cost on to its consumers in the form of higher prices while absorbing the remainder in the form of lower profits and decreased sales. In this way, the firm and the users of its output pay for the reductions in pollution caused by the firm's production. This appears to be an equitable solution. But even here, government makes some indirect contribution to financing the firm's control efforts. The cost of control can be written off as an expense for tax purposes and thus reduces accounting profits and income tax.

One major argument often used for not having the firm pay the entire cost of control is that the firm may be on the brink of economic disaster ("this plant is already losing money"), and any additional cost will cause it to close and move elsewhere. This in turn would cause an economic hardship on the community: Employment would be cut. Retail sales would be down. Property-tax collections would be reduced without a corresponding reduction in the local government's expenses. This is the typical threat. Perhaps the most rational counterargument for this is that if the firm (plant) is really on the brink, not causing it to pay for air pollution controls will only postpone the closing for a short time. If the damages of its pollution are large, perhaps the community as a whole would be better off without the employment and the pollution. Some communities now realize continued growth is not always good. In fact, the State of Oregon now operates under a quasi-official policy of not trying to attract new people to the state. Also, this closing could free resources in the community for the development of "clean" industries or firms more willing or able to control their pollution.

There is also an argument in favor of having the recipients of pollution pay for its control. When pollution is controlled, the recipients benefit from less pollution. It seems only right that they should pay for the benefits which they receive. This appears to be a less convincing argument on ethical grounds. However, it does have political appeal because it allows for a small charge to each of many recipients rather than a large charge to a relatively small number of polluters. The political advantages of this should be obvious.

One of the real problems of this approach is to determine who the recipients are and how much each should pay. Theoretically, all recipients who benefit from control should pay. But the dividing line is always fuzzy on such things. For example, should a person who lives outside the affected area but works within it pay? Perhaps so, but what if he only comes into the affected area once a week or once a month? The decision on how each recipent should pay is equally unclear. We could adopt the benefit principle and say that each recipient should pay in proportion to the benefits he receives. But then some meaningful measure of benefits must be derived. The measurement of such benefits has been one of the most difficult tasks that economists have undertaken, and the techniques available are imprecise and incomplete. For example, suppose that two people received the same level of improvement in air quality. One of the two people is in good health, but the other suffers from emphysema. Improved-air quality is more important to the emphysema sufferer, and we would conclude from the benefits-received principle that he should pay more for control. In this example, society tells the individual: "Don't get sick, or we will charge you more for air pollution control." This hardly seems like the most equitable position to take.

With these conflicts and problems, the natural reaction is to turn to government for a solution. The problem of air quality is important to all members of our society. Therefore, the argument goes, all members should help financially with its solution. So government supports research and control efforts and sometimes subsidizes pollution-control efforts of individual polluters. This financial help may consist of direct grants to municipalities or firms and/or investment credits or property-tax exclusions for privately installed control devices.

CONCLUSION

As with the other alternatives, there are problems with government subsidies. Such financial support will most often come from the general funds. The taxes that supply revenue to the general funds are not very good indicators of the taxpayer's ability to pay. This is increasingly true as one analyzes tax burdens among income groups in successively more local jurisdictions. Income, purchases, and property-improvement value, but certainly not land value, are also generally poor reflectors of the benefits received from control. Thus this form of financing does not fare well on equity grounds. Further, subsidies typically take a form that reduces the cost of control devices but does not affect the cost of other alternatives such as process adjustments and fuel conversion. This provides a bias toward devices and a possible misallocation of resources. Also, it is likely that some polluters will receive subsidies on divices which they would have installed regardless. This means that not all of the government's subsidy is stimulating control efforts and thus involves waste of very scarce control resources.

We can say that the question of who pays has no single, "best" solution, and like the control-policy alternatives, compromises will have to be made between competing goals. In this case, the competing goals are equity, political expediency, and sufficiency of revenue. Thus, it should be clear from the above discussion, that the decision as to who pays for air pollution control is not independent of the cost and effectiveness of alternative controls, nor is it independent of the control instruments employed.

The final consideration for all aspects of administrative-structure decisions is public acceptance of the decisions. If the public accepts and actively supports control activities, enforcement will be easier, and financing will be available. If the control effort lacks public support, it is likely to be ineffective.

INTERRELATIONSHIPS AMONG COMPONENTS OF THE MODEL

As has been suggested before, many of the components of this system are not completely independent of one another. In what follows and in Figure 1, only a few of the possible interrelationships are spelled out. There are many others, but space limitations and the likely loss of the central core in a maze of lines prevent their full exploration.

Let us suppose, for example, that used and new cars were stringently controlled so that many cars employed expensive control devices. This would increase the cost of individual automobiles relative to other forms of transportation. This in turn would probably result in increased use of rapid transit; additional electric power would then be consumed which would cause an increase in the output of electric power and consequently of air pollution. The increased cost of transportation would likely cause a shift in land utilization so that individuals would tend to live closer to where they worked. This reduces the total number of miles traveled during the day and reduces total emissions from cars and mass transportation. Such an effect is dependent on the ability to change one's place of residence and on the kind of control instrument employed.

If a regulation were employed that required the installation and maintenance of this expensive device regardless of the anticipated use of the car and the maintenance cost did not increase significantly with use, after the individual purchased the device he would not consider the cost of the air pollution he released (no device is 100 percent effective) in his decision to drive. In fact, he might be stimulated to make maximum use of his car in order to spread the cost over as many miles as possible. This type of regulation is likely to reduce the number of cars on the road but may increase the number of miles driven per car. Furthermore, if the individual must pay for the device, the low-income individual will be more likely to forego ownership of an automobile. This has socially undesirable implications in terms of finding and commuting to a place of employment. This in part can be offset by the development of a satisfactory mass transit system. The increased cost of transportation and the disruption of the social structure implicit in the shifting land use pattern may disuade people from locating in the airshed, but the resultant increase in air quality may prove to be an attracting force. The result of this complexity of relationships is obviously unknown. Research in this area may prove most helpful to decisionmakers.

THE DESIRED LEVEL OF QUANTIFICATION OF THE MODEL

The above model may be useful in discussing and thinking about the issues involved in air pollution control, but to what extent should it be improved, implemented, and quantified? The answer to this question depends on the extent to which the results of the quantified model effect air pollution control decisions. We have assumed that the model plays a central role in the policy-formulation process. In such a case, the model would be quantified to the point where an additional dollar spent on quantification would equal the additional benefit in terms of costs of control saved and/or improvements in air quality. If, however, the results of model calculations are ignored by decisionmakers, then quantification is useless except to the extent

CONCLUSION 185

that publicizing the results may put pressure on the decisionmakers to include this information in their deliberations.

Finally, let us discuss the practical problems involved in the implementation of such a model, given current knowledge. None of the submodels conceptualized in Figure 1 has as yet been developed. Let us take some of the principle components of the least cost emission control sets section as examples.

The Economic and Regional Development Model would be an attempt to interrelate the economic activity of the region to the people and resources of the region, its land-use pattern, its transportation network, and its social overhead capital network. It would also try to predict the effect of changes in the structure of the regional economy on each of these components and on the emissions released. Furthermore, it would attempt to estimate the feedback effects of control efforts on the regional economy. Such a model has not been successfully developed to date. It would be extremely complex, its quantification would be expensive, and it would often involve the collection of basic data. This is not to suggest, however, that such an effort is impossible. The many transportation-planning models that have been developed under federal grants to regional-planning commissions are steps in this direction. Furthermore, new models being constructed by the National Bureau of Economic Research and Resources for the Future show promise of furthering our understanding of these relationships.

The Air Chemistry and Meteorology Simulation Model is also untried. Some efforts have been made by Behar (1970), and others, but progress is slow and expensive. This model requires an understanding of all the major sources of pollution in an airshed as well as the chemical reactions and meteorological conditions that translate these emissions into various air qualities. The whole system is very incompletely understood.

Much of the basic data required to implement both the economics submodel and the meteorology submodel is not now available. The Emission Sources and Outputs inventory has never been thoroughly thought out, let alone taken. In all the research to date, data has been one of the major stumbling blocks. In almost all cases, control officials have only a very general idea as to the emissions of their jurisdiction. This condition may have been adequate for past efforts, but increasing concern with air quality is likely to indicate a desire for better quantification of source emissions. This effort will be expensive, time consuming, and, in some cases, not technically feasible in the next few years.

On the benefit side, most of the data on damages that is presently available is not in a form which allows it to be used in the suggested analysis. Especially in the human-health area, air pollutant effects are usually put in terms of thresholds. Thus, we speak of odor-perception thresholds, eye-irritation thresholds, or "significant" lung disease effects. Our analysis asks questions in incremental terms. For example, we would like to have an estimate of how many fewer persons suffer from lung disease caused by an air pollutant if its average or peak level is reduced by one unit (or two units, three units, etc.), over a specified period of time. In a few cases, some answers of this sort can be obtained from a simple reformulation of the existing data. In other cases, additional research will be required to provide the answers.

In the government decisionmaking and administration submodel, there are also many areas where much additional work would be necessary to fully understand the system. Because these areas are the major topics of the individual chapters, they will not be discussed here.

In summary, a policy trade-off analysis model such as is advocated here would be difficult and expensive to implement. However, we feel that the expected benefits in terms of understanding the problem and its complexity more thoroughly will justify some expenditure of funds.

**AN AIR POLLUTION
BIBLIOGRAPHY
FOR THE SOCIAL
SCIENCES**
Michael L. Fox

ECONOMICS

"AIPE Survey of Air Pollution Costs," Modern Manufacturing, 1:1 (June, 1968), 186-88.

"Air and Water Pollution: Does it Limit Industrial Expansion?" Industrial Development, 136 (July-August, 1967), 14-21.

Air Conservation. Part 3, Chapter VI. Washington, D.C.: Publication No. 80, American Association for the Advancement of Science, 1965.

"Air Pollution and Industrial Development," American Forests, 74 (January, 1968), 5.

Anderson, R. J., Jr. A Note on Economic Base Studies and Regional Econometric Forecasting Models. Lafayette: Department of Economics, Purdue University, 1969.

_____. "Application of Engineering Analysis of Production to Econometric Models of the Firm." Unpublished Ph.D. dissertation, University of Pennsylvania. Summarized in American Economic Review, 60 (May, 1970), 1969.

_____, and T. D. Crocker. "Air Pollution and Housing: Some Findings." Paper No. 264. Institute for Research in the Behavioral, Economic, and Management Sciences. Lafayette, Ind.: Krannert Graduate School of Industrial Administration, Purdue University (January, 1970).

_____. "The Site Value Approach to the Measurement of Economic

---

Some of the titles included were gleaned from the following sources: (a) The National Air Pollution Control Administration, Office of Technical Information and Publications (Raleigh, North Carolina), (b) Brian W. Peckham, "Recent Literature on the Economics of Air Pollution." Raleigh, North Carolina: Division of Economic Effects Research, National Air Pollution Control Administration,) October, 1969; and (c) bibliographic references presented with each of the literature surveys by contributors of this volume.

Losses Due to Air Pollution." New York: Paper presented at Annual Meeting of the Air Pollution Control Association, June 22-26, 1969.

"Antipollution Outlays to Rise 34 Percent," Engineering News-Record, 180:18 (May 2, 1968), 127.

Armour Research Foundation. The Economic Cost of Air Pollution. Chicago: Illinois Institute of Technology, ARF Project, No. 9-851, 1960.

Arrow, K. J., and G. Debreu. "Existence of an Equilibrium for a Competitive Economy," Econometrica, 22 (July, 1954) 265-90.

Ayres, Robert U. "Air Pollution in Cities," National Resources Journal, 9 (January, 1969), 1.

———, and A. V. Kneese. "Production, Consumption, and Externalities," American Economic Review, 59 (June, 1969), 282-97.

Barber, J. C. "Air Pollution: The Cost of Pollution Control," Chemical Engineering Progress, 64 (1968), 78.

Barnett, H. J., and C. Morse. Scarcity and Growth. Baltimore: Johns Hopkins University Press, 1963.

Bates, D. V. "Health Costs of Diseases Related to Air Pollution." Pollution and Our Environment. Conference Background Papers, Paper A4-2-4, Vol. 1. Montreal: Canadian Council of Resource Ministers, 1967.

Battelk Memorial Institute. Study of Unconventional Thermal, Mechanical, and Nuclear Low-Pollution Potential Power Sources for Urban Vehicles. Raleigh, N.C.: Prepared under Contract No. PH 86-67-109 for the National Air Pollution Control Administration (October, 1969).

Baumol, W., and R. Quandt. "Rules of Thumb and Optimally Imperfect Decisions," American Economic Review (March, 1964), 23-46.

Bechtel Corporation. "The Economics of Residual Fuel Oil Desulfurization." A study for the Division of Air Pollution, U.S. Public Health Service, Job 4728 (June, 1964).

Bell, M. "Cost Estimating: The Cost of Clean Air," Air Conditioning, Heating, and Ventilating, 65:7 (July, 1968) 41-46.

Blalock, H., and A. Blalock. "Towards a Clarification of Systems Analysis in the Social Sciences," Philosophy of Science, 26 (1959), 84-92.

Boulding, K. E. "The Economics of the Coming Spaceship Earth." Environmental Quality in a Growing Economy. H. Jarrett, editor. Baltimore: John Hopkins University Press, 1969, pp. 3-14.

_____. "The Economist and the Engineer: Economic Dynamics of Water Resource Development." Economics and Public Policy in Water Resource Development. Stephan C. Smith and Emery N. Castle, editors. Ames: Iowa State University Press, 1967, pp. 123-39.

Breton, A. "Towards an Economic Theory of Pollution Control and Abatement." Pollution and Our Environment. Conference Background Papers, Paper D28-1. Montreal: Canadian Council of Resource Ministers, 1967.

Brooks, Douglas L. "Industrial Development and Pollution Control: Friends or Foes?" AIDC Journal, 3:3 (1968), 1-13.

Brown, M. On the Theory and Measurement of Technological Change. New York: Cambridge University Press, 1966.

Buchanan, James M. "A Behavioral Theory of Pollution," Western Economic Journal, 6 (1968).

_____. "External Diseconomies, Corrective Taxes, and Market Structure," American Economic Review, 59 (March, 1969), 174-77.

_____, and G. Tullock. The Calculus of Consent. Ann Arbor: University of Michigan Press, 1962.

Burton, Ellison, and William Sanjour. "Multiple Source Analysis of Air Pollution Abatement Strategies," Federal Accountant, 18 (March, 1969), 48-69.

_____, and Sam R. Peterkin. A Cost Effectiveness Approach to Urban Air Pollution Abatement. Presented at the Joint National Meeting of the Operations Research Society of America and the Institute of Management Sciences, San Francisco, May 3, 1968. APTIC 1103. Washington, D.C.: Ernst and Ernst.

Carlson, J. W. "What Price a Quality Environment," Soil, Water Conservation Journal, 24:3 (May-June, 1969), 84-88.

Carter, Peter. "Industry's Struggle Against Air and Water Pollution," New Jersey Business, 12 (September, 1965), 19-23.

Cass, J. S. "Flourides: A Critical Review IV," Journal of Occupational Medicine, 34 (October and November, 1961) 471-77 and 527-43.

Cheung, S. N. S. "Transactions Costs, Risk Aversion, and the Choice of Contractual Arrangements," Journal of Law and Economics, 12 (April, 1969), 23-47.

Coase, R. "The Problem of Social Cost," Journal of Law and Economics (October, 1960), 1-43.

Control Techniques for Particulate Air Pollutants. National Air Pollution Control Administration. Pub. No. AP-51. Chapter Six: "Economic Consideration in Air Pollution Control." Washington D.C., June, 1969.

Crane, D. "New Strategy in the War Against Filthy Air," Financial Post, 59 (June 19, 1965), 15-16.

Crawford, W. D. "The Cost of Clean Energy," Air Pollution Control Association Journal, 19:5 (1969), 322-24.

Crocker, B. B. "Minimizing Air Pollution Control Costs," Chemical Engineering Progress, 64, April 1968, 79-86.

Crocker, T. D. "Externalities, Property Rights, and Transactions Costs: An Empirical Study," Journal of Law and Economics (forthcoming), 1971.

_____. "On Air Pollution Control Instruments," Hastings Law Journal (February, 1971).

_____. "Some Economic Aspects of Air Pollution Control with Special Reference to Polk County, Florida." Report to the U.S. Public Health Service, (January, 1968).

_____. "Some Economics of Air Pollution Control," Natural Resources Journal, 8 (April, 1968), 236-58.

_____. "Structuring of Atmospheric Pollution Control Systems." The Economics of Air Pollution. H. Wolozin, editor. New York: W. W. Norton, 1966, pp. 61-86.

_____. "The Economics of Environmental Quality." Paper presented at the Seminar on National Economic Policy, American University, Washington, D.C., 1966.

_____. "The Efficacy of Alternative Means of Pollution Control," Proceedings, Committee on Pollution, National Academy of Sciences--National Research Council, Washington, D.C., 1965.

_____. "The Measurement of Economic Losses from Encompensated Externalities." Economics of Air and Water Pollution. W. R. Walker, editor. Blacksburg, Va.: Virginia Water Resources Center, Virginia Polytechnic Institute, 1969, pp. 223-39.

_____, and Robert J. Anderson, Jr. "The Site Value Approach to the Measurement of Economic Losses Due to Air Pollution." Paper presented at the Annual Meeting, Air Pollution Control Association, New York, June, 1969.

_____, and A. J. Rogers III. Environmental Economics. Hindsdale, Ill.: Dryden Press, 1971.

Dales, J. H. Pollution, Property and Prices: An Essay in Policy Making and Economics. Toronto: University of Toronto Press, 1968.

Danø, S. Industrial Production Models. New York: Springer-Verlag, 1966.

d'Arge, Ralph, Truman Clark, and Osman Bubik. "Automotive Exhaust Emissions Taxes: Methodology and Some Preliminary Tests," Research Project S-12, Project Clean Air, University of California Research Reports, Vol. 3. Riverside, 1970.

Davis, O., and A. Whinston. "Externalities, Welfare, and the Theory of Games," Journal of Political Economy, 70 (June, 1962), 241-62.

Debreu, G. Theory of Value. New York: Wiley, 1959.

Degler, S. E. "State Air Pollution Control Laws." Environmental Management Series. Washington, D.C.: Bureau of National Affairs, 1969.

Demetz, H. "The Exchange and Enforcement of Property Rights," Journal of Law and Economics, 7 (October, 1964), 11-31.

Dennis, C. "How Much Will Pollution Control Cost You?" Electric Light Power (June, 1969), 84-85.

Dolbear, F. T., Jr. "On the Theory of Optimum Externality," American Economic Review, 57 (March, 1967) 90-103.

Downing, Paul B., and Lytton Stoddard. "Benefit/Cost Analysis of Air Pollution Control Devices for Used Cars," Research Project S-10, Project Clean Air, University of California Research Reports, Vol. 3. Riverside, 1970.

Downs, A. An Economic Theory of Democracy. New York: Harper & Row, 1957.

Edmisten, Norman G., and Francis L. Bunyard. "A Systematic Procedure for Determining the Cost of Controlling Particulate Emissions from Industrial Sources." Paper presented at the Annual Meetings, Air Pollution Control Association, June, 1969.

Erlich, P. The Population Bomb. New York: Ballentine Books, 1969.

Ernst and Ernst, Washington, D.C. A Cost-Effectiveness Study of Air Pollution Abatement in the Kansas City Area. Presented at the Kansas City Air Pollution Abatement Conference, Kansas City, Mo., 1968. APTIC 11111.

_____. "A Cost-Effectiveness Study of Air Pollution Abatement in the National Capital Area." A report to the U.S. Public Health Service for Contract No. PH 86-68-37, January, 1969.

_____. "A Cost-Effectiveness Study of Particulate and SO_x Emission Control in the New York Metropolitan Area." Presented at the Air Pollution Abatement Conference, New York, February, 1968. APTIC 11114.

_____. "Applications fo Cost-Effectiveness Analysis to Air Pollution Control." A report to the U.S. Public Health Service for Contract No. CPA 22-69-17, September, 1969.

_____. A Rapid Cost Estimating Method for Air Pollution Control Equipment. Final Report. PH-86-68-37, U.S. Public Health Service, Washington, D.C., 1968.

Faith, W. L. "Economics of Motor Vehicle Pollution Control," Chemical Engineering Progress, 62 (October, 1966), 41-43.

Faltermayer, Edmund K. "We Can Afford Clean Air: Polluted Air Is Corroding Metals, Menacing Health and Degrading the Human Spirit; for Around $3 Billion a Year," Fortune, 72, November, 1965), 158-65.

Federal Coordinating Committee on the Economic Impact of Pollution Abatement, Working Committee on Secondary Impact of Air Pollution Abatement, Second Report, Washington, D.C., December 15, 1967. APTIC 11185.

_____. Working Committee on Economic Incentives. "Cost Sharing with Industry?" Summary Report. Washington, D.C., November 20, 1967. APTIC 1190. 37.

Fink, K. "Economics of Air Pollution," Smokeless Air, 82 (Autumn, 1968).

Flagg, Samuel B. "The Dividends that Float Up the Chimney: What Smoke Means in Dollars and Cents," Scientific American, 120 (1912), 539.

Fogel, M. E., et al. "Comprehensive Economic Cost Study of Air Pollution Control Costs for Selected Industries and Selected Regions." Research Triangle Park, N.C.: Research Triangle Institute for NAPCA Contract No. CPA-22-69-79, February, 1970.

Frankel, R. J. "Economic Impact of Air and Water Pollution Control on Coal Preparation," Mining Congress Journal, 54 (October, 1968), 56-64.

_____. "Problems of Meeting Multiple Air Quality Objectives for Coal-Fired Utility Boilers," Journal of the Air Pollution Control Association, 19 (January, 1969), 18-23.

Freeman, A. M., III. "The Distribution of Environmental Quality." A paper read before the RFF Conference on Research in Environmental Quality, June 16-18, 1970.

Frisch, R. "On Welfare Theory and Pareto Regions," International Economic Papers, 9 (1959), 39-92.

Gaffney, M. M. "Welfare Economics and the Environment." Environmental Quality in a Growing Economy. H. Jarret, ed. Baltimore: Johns Hopkins University Press, 1966, pp. 88-101.

Galbraith, J. K. "Economics and Environment," American Institute of Architects Journal, 46 (September, 1966), 55.

Gerhardt, Paul. "Financial Incentive to Air Pollution Control." Paper presented at the Sixty-second annual meeting, Air Pollution Control Association, New York, June 22-26, 1969.

Gibson, W. B. "Economics of Air Pollution." Proceedings of the First National Pollution Symposium, Stanford Research Institute, November, 1949.

Goldman, Marshall I., ed. Controlling Pollution: The Economics of a Cleaner America. Englewood Cliffs, N.J.: Prentice-Hall, 1967.

Goldner, L. "Air Pollution in the Metropolitan Boston Area." The Economics of Air Pollution. H. Wolozin, editor. New York: W. W. Morton, 1966, pp. 127-61.

Gordon, H. S. "The Economic Theory of a Common Property Resource: The Fishery," Journal of Political Economy, 62 (April, 1954), 124-42.

Graef, J. de V. Theoretical Welfare Economics. New York: Cambridge University Press, 1967.

Gramm, William P. "A Theoretical Note on the Capacity of the Market System to Abate Pollution," Land Economics (August, 1969).

Green, H. A. G. Aggregation in Economic Analysis. Princeton: Princeton University Press, 1964.

Grosse, Robert N. "Some Problems in Economic Analysis of Environmental Policy Choices." Proceedings of the Symposium of Human Ecology, Arlie House, Warrenton, Virginia, November, 1968. Washington, D.C.: U.S. Public Health Service CPE, 1969.

Gustavson, Reuben G. "What Are the Costs to Society?" Proceedings of the National Conference on Air Pollution, November 1958. Washington, D.C.: Public Health Service, Pub. 654, 1959.

Gutmanis, Ivars, and L. Goldner. "Welfare Economics and Public Policy: Parallels in the Analysis of the Impact of Air Pollution and Weather Modification." Paper presented at the Symposium on the Economic and Social Aspects of Weather Modification, Boulder, Colorado, July 1-3, 1965.

Hamburg, F. C. "Economically Feasible Alternatives to Open Burning in Railroad Freight Car Dismantling," Journal of the Air Pollution Control Association, 19 (July, 1969), 477-83.

Hanks, James J., and Harold D. Kube. "Industry Action to Combat Pollution," Harvard Business Review, 44 (September-October, 1966), 49.

Harris, L. E., et al. "Effects of Flourine on Dairy Cattle," Journal of Animal Science, 23 (May, 1964) 537-45.

Havighurst, C. C. "A Survey of Air Pollution Litigation in the Philadelphia Area." Durham, N.C.: Duke University School of Law for NAPCA, 1970. Contract No. CPA 22-69-112.

Heller, A. N. "Methods of Evaluating Socioeconomic Effects of Air Pollution." Proceedings of the International Conference on Atmospheric Emissions from Sulfate Pulping, Sanebel Island, Florida, April 28, 1966. E. R. Hendrickson, ed. National Council for Stream Improvement and the University of Florida, 1966, pp. 141-56.

Hepting, G. H. "Damage to Forests from Air Pollution," Journal of Forestry 62 (1964), 630-35.

Herfindahl, Orris C. Natural Resources Information Economic Development. Baltimore: John Hopkins University Press, 1969.

―――――, and Allen V. Kneese. Quality of the Environment: An Economic Approach to Some Problems in Using Land, Water and Air. Baltimore: John Hopkins University Press, 1965.

Hicks, J. R. "The Four Consumer Surpluses," Review of Economic Studies, 11 (1943), 31-41.

Higgens, R. J. "Two Cities, Milford and Orange, Connecticut, Jointly Fight Air Pollution . . . and Find, That by Working Together, They Are More Effective and the Work is Less Costly," American City (June, 1968), 137-38.

Hirshleifer, J. "Investment Decision Under Uncertainty: Choice-Theoretical Approaches," Quarterly Journal of Economics, 79 (November, 1965), 509-36.

Howard, William C. "A New and Economic Solution to the Problems of Stream and Air Pollution," Norsk Skogind, 21 (April, 1967). APTIC 7769.

Huey, Norman A. "Economic Benefits from Air Pollution Control." Cincinnati: National Center for Air Pollution Control, U.S. Public Health Service, 1967. APTIC 9102.

Jackson, W. E., H. C. Wohlers, and W. DeCoursey. "Determining Air Pollution Control Costs," Journal of the Air Pollution Control Association, 19: 12 (December, 1969), 917-23.

Jensen, D. A. "Status Report on Cost Factors in Exhaust Control," Journal of the Air Pollution Control Association, 14 (October, 1964), 532-36.

Jimeson, R. M. "The Possibilities of Solvent Refined Coal." Unpublished master's thesis, George Washington University, February, 1965.

Johnson, Hal. "The High Cost of Foul Air," Progressive Farmer (April, 1968).

Kafoglis, Milton Z. "The Economics of Environmental Engineering," Engineering Progress, 21:6 (1967), 72. Bulletin No. 128, Water Resources Research Center, University of Florida.

Katell, S., and K. D. Plants. "Here's What SO_2 Removal Costs," Hydrocarbon Processing, 46 (July, 1967), 161-64.

Klein, L. R. An Introduction to Econometrics. Princeton: Prentice-Hall, Inc., 1962.

Kneese, A. V., and d'Arge, R. C. "Pervasive External Costs and the Response of Society." U. S. Congress, Joint Economic Committee, The Analysis and Evaluation of Public Expenditures, Part I. Washington, D.C.: 91st Congress, 1st Session, 87-117.

Knetsch, Jack. "Economic Aspects of Environmental Pollution," Journal of Farm Economics, 48 (1969), 5.

Kohn, Robert E. "Achieving Air Quality Goals at Minimum Cost," Washington University Law Quarterly (Spring, 1968), 325.

_____. Economic Criteria for Air Pollution Control: A Case Study of Missouri Regulations of Sulfur Dioxide. St. Louis: Washington University, Department of Economics, June, 1967.

_____. "A Linear Programing Model for Air Pollution Control in the St. Louis Airshed." Unpublished Ph.D. dissertation, Washington University, 1968.

_____. "Joints-Outputs of Liquid, Land, and Thermal Wastes in a Linear Programing Model for Air Pollution Control." Paper presented at the American Statistical Association meeting, Detroit, December, 1970.

Koogler, J. B., et al. "A Multivariate Model for Atmospheric Dispersion Predictions," Journal of the Air Pollution Control Association, 17 (April, 1967), 211-14.

Krutilla, J. "Conservation Reconsidered," American Economic Review (September, 1967), 77-86.

Laffer, W. G. "Industry and Pollution: A Businessman Urges Companies to Lead Fight on Problem," Wall Street Journal, 166 (November 26, 1965), 10.

Lancaster, K. J. "Change and Innovation in the Technology of Consumption," American Economic Review, 56 (May, 1966), 14-23.

_____. "A New Approach to Consumer Theory," Journal of Political Economics, 74, (April, 1966), 132-57.

Landau, Emanuel. "Economic Aspects of Air Pollution as It Relates to Agriculture," Agriculture and the Quality of Our Environment, No. 11. Washington, D.C.: AAAS Pub. 85, 1967.

Lave, L. B., and E. P. Seskin. "Air Pollution and Human Health: The Dollar Benefit of Pollution Abatement." Pittsburgh: Graduate School of Industrial Administration, Carnegie-Mellon University, October, 1969.

Leclerc, E. "Economic and Social Aspects of Air Pollution." Air Pollution. Geneva: World Health Organization, 1961.

Leclerc, M. "General Report: Economic Effects of Air Pollution," European Conference on Air Pollution, 1964. Strasbourg, France: Council of Europe, 1964, pp. 71-80.

Lind, R. C. "Land Market Equilibrium and the Measurement of Benefits from Urban Programs." Paper delivered at the CUE Conference, University of Chicago, September 11-12, 1970.

Liston, L. "Air and Water Pollution: Does It Limit Industrial Expansion?" Industrial Development, 136: 5 (July-August, 1967), 14-21.

Loveridge, R. O. See Political Science Section.

Maass, A. "Benefit-Cost Analysis--Its Relevance for Public Investment Decisions," Quarterly Journal of Economics (May, 1966), 208-26. "Comment," Quarterly Journal of Economics (November, 1967), 695-702.

Marshak, T., et al. Strategy for R and D. New York: Springer-Verlag, 1967.

McCaldin, R. O. "How Much Does Air Pollution Cost?" Air Engineering, 15: 18 (November, 1968).

McKee, H. C. "Preferential Tax Treatment for Pollution Control Expenditures: Engineering Consideration," Journal of the Air Pollution Control Association, 18: 9 (September, 1968), 596-99.

McKenzie, L. W. "On the Existence of a General Equilibrium for a Competitive Market," Econometrica, 27 (January, 1959), 54-71.

McLaughlin, J. F. "Atmospheric Pollution Considerations Affecting the Ultimate Capacity of a Thermal-Electric Power Plant Site," Journal of the Air Pollution Control Association, 17: 7 (July, 1967), 470-73.

Meckler, M. "Cost Estimating: Air Handling Equipment for Contamination Control," Air Conditioning, Heating, and Ventilation, 65: 7 (July, 1968), 37-40.

Medelia, N. A., et al. "Community Perception of Air Quality: An Opinion Survey in Clarkson, Washington." Washington, D.C.: U.S. Public Health Service, Pub. No. 999-AP-10, 1965.

Michelson, Irving. The Costs of Living in Polluted Air Versus the Costs of Controlling Air Pollution. New Rochelle, New York: Environmental Health and Safety Research Associates, 1966. APTIC 7907.

_____, and B. Tourin, "Comparative Method for Studying Costs of Air Pollution," Public Health Reports, 81: 6 (June, 1966), 505-11.

_____. The Household Costs of Living in Polluted Air in the Washington, D.C. Metropolitan Area. New Rochelle, New York: Environmental Health and Safety Research Associates, 1966. APTIC 9044.

_____. The Household Cost of Air Pollution in Connecticut. Preprint, New Rochelle, N.Y.: Environmental Health and Safety Research Associates, October 1, 1968. APTIC 11084.

Middleton, J. R. "Photochemical Air Pollution Damage to Plants," Annual Review of Plant Physiology, 12 (1961), 431-39.

Mills, E. S. "Economic Incentives in Air Pollution Control." The Economics of Air Pollution. H. Wolozin, editor. New York: W. W. Norton, 1966, pp. 40-55.

_____. "Federal Fiscal Policy in Air Pollution Control." Proceedings of the National Conference on Air Pollution, 3, 1966. PHS Pub. No. 1649, 574-78.

Minifie, J. M. "United States Rehabilitation Benefits Canada," Business Quarterly, 32 (Spring, 1967), 8, 86.

Mishan, E. J. The Costs of Economic Growth. New York: Praeger, 1967.

──────. "Pareto Optimality and the Law," Oxford Economic Papers, 19 (November, 1967), 255-87.

Morrison, C. C. "Generalizations on the Methodology of Second Best," Western Economics Journal, 6 (March, 1968), 112-20.

Murphy, E. F. Governing Nature. Chicago: Quadrangle Books, 1967.

Muth, R. F. "Economic Change and Rural-Urban Land Conversions," Econometrica, 29 (January, 1961), 1-23.

National Air Pollution Control Administration. "Control Techniques for Hydrocarbon and Organic Solvent Emissions from Stationary Sources." Washington, D.C.: U.S. Public Health Service, Pub. No. AP-68, March, 1970.

──────. The Cost of Clean Air, First Report of the Secretary of Health, Education, and Welfare to the Congress in Compliance with the Air Quality Act of 1967 (June, 1969), Washington, D.C.

──────. "Guidelines for the Development of Air Quality Standards and Implementation Plans." Washington, D.C.: U.S. Government Printing Office, May, 1969.

──────. "Guide to Research in Air Pollution." Raleigh, N.C.: NAPCA Pub. No. AP-47, U.S. Public Health Service, April, 1970.

──────. "Prospects for Electric Vehicles--Electric." Raleigh, N.C.: Arthur D. Little, Inc., October, 1969. Prepared under Contract No. PH 86-67-108.

Nelson, F., and L. Shenfield. "Economics, Engineering, and Air Pollution in the Design of Large Chimneys," Journal of the Air Pollution Control Association, 15 (August and November, 1965), 340-45 and 586-89.

"New Plants Invest Heavily in Pollution Control," Modern Manufacturing, 2: 5 (May, 1966), 306-8, 310.

Norsworthy, J. R., and A. Teller. "The Use of Input-Output Analysis to Design a Study of the Regional Impact of Air Pollution Control." Paper presented at the Sixty-second Annual Meeting, Air Pollution Control Association, New York, June 22-26, 1969.

Nourse, H. O. "The Effect of Air Pollution on House Values," Land Economics, 43 (May, 1967), 181-89.

_____. "Is There an Economic Solution to the Air Pollution Problem?" Bureau of Community Planning Newsletter (Fall, 1969).

Oakland, W. H. "Joint Goods," Economica (August, 1969), 253-68.

O'Connor, J., and Joseph F. Citarella. "An Air Pollution Control Cost Study of the Steam-Electric Power-Generating Industry," Journal of the Air Pollution Control Association, 20 (May, 1970), 283-88.

Ogden, Delbert C. "Economic Analysis of Air Pollution," Land Economics, 42 (May, 1966), 137.

Olatka, F. T., and J. N. Brogard. "The Economic Aspects of Engineering for Industrial Air Pollution Control." Paper presented at the Sixty-second Annual Meeting, Air Pollution Control Association, New York, June 22-26, 1969.

Ozolins, G., and R. Smith. "Rapid Survey Technique for Estimating Community Air Pollution Emissions." Washington, D.C.: U.S. Public Health Service, Pub. No. 999-AP-29.

Panel on Electrically Powered Vehicles. "The Automobile and Air Pollution: Part II." Washington, D.C.: U.S. Department of Commerce, December, 1967.

Park, W. R. "Air Pollution Economics: Justifying Industry's Investment," Consulting Engineer, 31: 12 (December, 1968), 40, 42, 44, 46

_____. Air Pollution Economics, the Cost of Control," Consulting Engineer, 32: 1 (January, 1969), 58, 61.

_____. "The Cost/Benefit Analysis," Consulting Engineer, (February, 1969), 112, 115.

Pearse, I. F. A Contribution To Demand Analysis. Oxford: Clarendon Press, 1964.

Pelle, W. J. Bibliography on the Planning Aspects of Air Pollution Control, Summary and Evaluation. Washington, D. C.: Public Health Service, 1964.

Perloff, H. S., and Lowdon Wingo, Jr. Issues in Urban Economics. Baltimore: Johns Hopkins University Press, 1968.

Peterson, J. T., "The Climate of Cities." Raleigh, N.C.: Natiional Air Pollution Control Administration, Pub. No. AP-59, October, 1969.

Plott, C. "Externalities and Corrective Taxes," Economica, 33 (1966), 84.

Prest, A. R., and R. Turvey. "Cost-Benefit Analysis: A Survey," Economic Journal (December, 1965), 683-735.

Proceedings of the National Conference on Air Pollution, November, 1958. Group D--Economic and Social Effects of Air Pollution, 241. Economic Aspects of Engineering Control . . . 314. Washington, D. C.: U.S. Public Health Service, Pub. No. 645, 1959.

Public Administration and Metropolitan Affairs Program, Southern Illinois University, Edwardsville. "Public Awareness and Concern with Air Pollution in the St. Louis Metropolitan Area." Washington, D.C.: U.S. Public Health Service, May, 1965.

Quirk, J., and R. Saposnik. Introduction to General Equilibrium Theory and Welfare Economics. New York: McGraw-Hill 1968.

Reinow, R., and L. Reinow. Moment in the Sun. New York: Ballentine Books, 1969.

"Resource Economics and a Quality Environment." Man Versus Environment. Monograph No. 3, National Sanitation Foundation, School of Public Health, University of Michigan, 1965.

Resources for the Future, Inc. Annual Reports, 1967, 1968, and 1969. Washington, D.C.: Resources for the Future, Inc.

Ridker, R. G. Economic Costs of Air Pollution, Studies in Measurement. New York: Praeger, 1967.

_____. The Problem of Estimating Total Costs of Air Pollution, A Discussion and an Illustration. Report of the U.S. Public Health Service, PH-86-64-2 and PH-86-65-17, July 1966.

_____, and L. A. Henning. "The Determinants of Residual Property Values with Special Reference to Air Pollution," Review of Economics and Statistics, 49 (1967), 246-57.

Rose, Sanford. "The Economics of Environmental Quality," Fortune, 81 (February, 1970), 120-23, 184-88.

Rydell, R. "Air Pollution and Urban Population Distribution." New York: Urban Research Center, Hunter College, July, 1968.

_____, and B. Stevens, "Air Pollution and the Shape of Urban Areas," Journal of the American Institute of Planners, 34 (January, 1968), 50-51.

Samuelson, P. A. "A Diagrammatic Exposition of a Theory of Public Expenditures," Review of Economic Statistics, 37 (November, 1955), 350-56.

_____. Foundations of Economic Analysis. Cambridge: Harvard University Press, 1947.

_____. "The Pure Theory of Public Expenditure," Review of Economics and Statistics, 36 (November, 1954), 387-89.

Schelling, T. C., et al. "The Life You Save May Be Your Own." Problems of Public Expenditure Analysis. S. Chase, Editor. Washington, D.C.: The Brookings Institution, 1968, pp. 127-76.

Scitovsky, T. "What Price Economic Progress," Yale Review (Autumn, 1959).

Scott, A. D., and J. D. Graham. "Economic Decisions about Air Pollution Control," Pollution and Our Environment, Conference Background Papers, 3. Montreal: Canadian Council of Resource Ministers, 1967. Paper D29-4, pp. 1-12.

Secretary, American Economic Association, "Report for the Year Ending December 31, 1968," American Economic Review, 59 (May, 1969), 578-83.

Semrau, K. T. "Dust Scrubber Design--A Critique on the State of the Art," Journal of the Air Pollution Control Association, 13 (December, 1963), 587-94.

Smith, R. G. "Economic Consideration in Air Pollution Control," Proceedings of the National Conference on Air Pollution, December, 1962. Washington, D.C., U. S. Public Health Service, Pub. 1022, 1963.

Smith, V. L. Investment and Production. Cambridge: Harvard University Press, 1966.

"Spending for Clean Air to Climb," Steel, 164:5 (February 3, 1969), 38-39

Stanley, W. J., and Peter A. Loguercio. "Economic Correlation of Air Pollution Potential with Index of Productivity and the GNP."

Paper presented at the sixty-second annual meeting of the Air Pollution Control Association, New York, June 22-26, 1969.

Starr, Roger, and James Carlson. "Pollution and Poverty: The Strategy of the Cross-Commitment," The Public Interest (Winter, 1968), 104-31.

Stern, A. C., ed. Air Pollution. 3 vols. New York: Academic Press, 1968.

Stockman, R. L., and Donald Anderson. "Physiologic, Economic and Nuisane Effects of Emissions from Sulfate Pulping." Proceedings of the International Conference on Atmospheric Emissions from Sulphate Pulping, Deland, Florida, April 28, 1966. National Council for Stream Improvement and the University of Florida, 1966.

Strodtbeck, F. L. "Professional Appraisers' Judgement of the Effect of Air Pollution on Property Values." Report of the U. S. Public Health Service, PH-86-76-44-Neg. 5, January, 1969.

Strotz, R. H. Economics of Urban Air Pollution. Washington, D.C.: Resources for the Future, 1966.

_____. "Use of Land Rent Changes to Measure Welfare Benefits of Land Improvement." Evanston, Ill.: Department of Economics, Northwestern University, July, 1967.

_____, and Colin Wright. A Price-Tax Solution to an Air Pollution Problem. Ottawa: Canadian Economic Association, June, 1967.

"Studies of the Structure of Economists' Salaries and Incomes," Committee on the National Science Foundation Report on the Economics Profession, American Economic Review, Part 2, 58 (December, 1968).

Swanson, E. R. "Problems of Applying Commercial Results to Commercial Practice," Journal of Farm Economics, 39 (May, 1957), 382-89.

Teller, A. "Air Pollution Abatement: An Economic Study into the Costs of Control." Unpublished Ph.D. dissertation, Johns Hopkins University, 1967.

_____. "Air Pollution Abatement: Economic Rationality and Reality," Daedalus, 96:4 (Fall, 1967), 1082-98.

_____. "The Use of Linear Programing to Estimate the Cost

of Some Alternative Air Pollution Abatement Policies." Proceedings of the IBM Scientific Computing Symposium on Water and Air Resource Management, Yorktown Heights, New York, October 23-25, 1967.

_____, and R. Norsworthy, Economic Aspects of Air Pollution. Philadelphia: Government Studies Center, Fels Institute of Local and State Government, University of Pennsylvania, 1969.

_____. "Regional Impact of Alternative Air Pollution Policies." Paper delivered at the Annual Air Pollution Control Association Meeting, New York, June, 1969.

Teworte, I. W. "Economic Aspects of Recovery of Mineral From Effluents," Chemistry and Industry, 18 (May, 1969), 565.

"The Cost of Clean Air," First Report of the Secretary of Health, Education, and Welfare Congress in Compliance with the Air Quality Act of 1967, June, 1969.

"The Economic Cost of Air Pollution." Chicago: Armor Research Foundation, Illinois Institute of Technology, Project No. 9-851, 1960.

"The Economic Cost of the Smoke Nuisance to Pittsburgh." Pittsburgh: Mellon Institute of Industrial Relations and School of Specific Industries, Bulletin No. 4, 1913.

"The Economics of Environmental Management--Panel Discussion," American Economic Review, 58 (May, 1968).

"The Structure of Economists' Employment and Salaries, 1964," Committee on the National Science Foundation, Report on the Economics Profession, American Economic Review, Part 2, 55, (December, 1965).

Theil, H. Economics and Information Theory. Amsterdam: North-Holland Publishing Co., 1968.

Thomann, R. V. "Mathematical Model for Dissolved Oxygen," Journal of Sanitary Engineering, Division of the American Society of Civil Engineers, 89: SA5 (October, 1968).

_____, and D. Marks. "Results from a Systems Analysis Approach to the Optium Control of Estuarine Water Quality." Paper presented at the Third International Conference on Water Pollution Research, Munich, September, 1966.

Thompson, Wilbur R. A Preface to Urban Economics. Baltimore: Johns Hopkins University Press, 1965.

Turner, D. B. "A Diffusion Model for an Urban Area," Journal of Applied Meteorology, 3, (February, 1964) 83-91.

Uhlig, H. H. "The Cost of Corrosion to the United States," Corrosion, 6 (January, 1950) 87-98.

"Union Officials Learn About Air Pollution," Environmental Science and Technology 3:5 (May, 1969), 429-30.

Verner, W. C., and M. S. Edsall. "Analysis of Economic Factors in Catalytic Air Treatment," Air Engineering, 38:9 (October, 1965), 41-50.

Vickery, W. "Theoretical and Practical Possibilities and Limitations of a Market Mechanism Approach to Air Pollution Control." Paper presented at the Annual Meeting of the Air Pollution Control Association, Cleveland, Ohio, June, 1967.

Walker, W., ed. Economics of Air and Water Pollution. Blacksburg, Va.: Virginia Water Resources Research Center, Virginia Polytechnic Institute, 1969.

Ward, B. N. The Socialist Economy. New York: Random House, 1967.

Watson, J. H. "Capital Expenditures for Pollution Abatement," Conference Board Record, 4 (September, 1967), 27-30.

Weber, H. J. "The Impact of Air Pollution Laws on the Small Foundry", Journal of the Air Pollution Control Association, 20 (February, 1970), 67-71.

Weidenbaum, M. L. "Federal Budgeting: The Choice of Government Programs." Washington, D.C.: American Enterprise Institute for Public Policy Research, 1964.

Weinrich, J. E. "Strategies for Controlling Air Pollution," Public Utilities Fortnightly, 81 (June 6, 1968), 80-84.

Weisbrod, B. "Income Redistribution Effects and Benefit-Cost Analysis." Problems in Public Expenditure Analysis. S. B. Chase, ed. Washington, D.C.: Brookings Institution, 1968, pp. 177-222.

Willett, H. P. "Cutting Air Pollution Control Costs," Chemical Engineering Progress, 63:3 (March, 1967), 80-83.

Williams, J. D. "Cost of Air Pollution and Pollution Abatement." Paper presented at the Regional Conference on Problems of Air Pollution, Washington University, St. Louis, Missouri, November, 1966.

Williamson, I., ed. "Hierarchical Control and Optimum Firm Size," Journal of Political Economics, 75 (April, 1967) 123-38.

Wilson, R. D., and D. W. Minnotte. "A Cost-Benefit Approach to Air Pollution Control," Journal of the Air Pollution Control Association, 19:5 (May, 1969), 303-8.

_____. "Control Costs vs. Benefits." Paper presented before the Kansas City, Missouri, Abatement Activity, Cincinnati, July, 1968.

_____. "Economics Impact of Air Pollution Control in the Ironton, Ohio-Ashland, Kentucky, Huntington, West Virginia Area." Paper presented at the Ironton-Ashland-Huntington Air Pollution Abatement Activity, Cincinnati, July, 1968.

_____. "Government Cost Sharing with Industry for Air Pollution Control," Journal of the Air Pollution Control Association, 19:10 (October, 1969), 761-66.

Winch, D. M. "Consumer's Surplus and the Compensation Principle," American Economic Review, 55 (June, 1965) 395-423.

Winkelstein, W., Jr., and S. Kanton. "The Relationship of Air Pollution and Economic Status to Total Mortality, a Selected Respiratory Mortality in Men," Archives of Environmental Health, 14 (January, 1967), 163-71.

Wolozin, H. "Economics of Air Pollution: Central Problems," Law and Contemporary Problems, 33:2 (Spring, 1968), 227-38.

_____. Economics of Air Pollution. New York: W. W. Norton, 1966.

_____. "Intrasigent Economic Behavior in Air Pollution Control and Decision." Paper presented at the Annual Meeting of the Air Pollution Control Association, Cleveland, June, 1966.

_____, and E. Landau. "Crop Damage from Sulfur Dioxide," Journal of Farm Economics, 48 (May, 1966), 394-402.

Woodcock, V. R., and Larry Barrett. "Economic Indicators of the Impact of Air Pollution Control: Gray-Iron Foundaries, A Case

Study," *Journal of the Air Pollution Control Association*, 20 (February, 1970), 72-78.

Wright, C. "Economic and Political Aspects of Dynamic Pollution Control." Paper presented at the Oregon State University Air Pollution Effects Seminar, October 23, 1969.

Wright, Colin. "Some Aspects of the Use of Corrective Taxes for Controlling Air Pollution Emissions," *Natural Resources Journal*, 9 (January, 1969), 63-82.

Yocum, J. E. "The Cost of Air Pollution--Effects and Control," *Tennessee Industrial Hygiene News*, 25 (1968), 1.

Zeckhauser, R. "Uncertainty and the Need for Collective Action." U. S. Congress, Joint Economic Committee. *Analysis and Evaluation of Public Expenditures*, Part I. Washington, D.C.: 91st Congress, 1st Session, 1969, pp. 149-66.

Zerbe, Richard O., Jr. *The Economics of Air Pollution: A Cost-Benefit Approach*. Report to the Department of Public Health, Government of the Province of Ontario. Toronto, July, 1969.

LAW

Aborn, R., and Carl E. Axelrod. "State Air Pollution Control Legislation," *Industrial and Commercial Law Review*, 9 (Spring, 1968), 217-56.

"A Compilation of Selected Air Pollution Emission Control Regulations and Ordinances." U.S. Department of Health, Education, and Welfare. Public Health Service, Pub. No. 999-AP-43, rev. ed., 1968.

"A Digest of State Air Pollution Control Laws," U. S. Department of Health, Education, and Welfare. Public Health Service, Pub. No. 711, 1967.

Adinolfi, G. "First Steps Toward European Cooperation in Reducing Air Pollution--Activities of the Council of Europe," *Law and Contemporary Problems*, 33:2 (Spring, 1968), 421-26.

Air Conservation Commission. "Air Conservation." American Association for the Advancement of Science, 1965.

"Air Pollution--Automobile Smog: A Proposed Remedy," DePaul Law Review, 14 (Spring-Summer, 1965), 436.

"Air Pollution Control," Law and Contemporary Problems, 33 (Spring, 1968), 195-426.

"Air Pollution Control: Enabling Legislation," American County Government, 31 (September, 1966), 31-46.

"Air Pollution Control: Inspection and Enforcement," American County Government, 31 (October, 1966), 27-42.

"Air Pollution Control in Minnesota," Minnesota Law Review, 54:953-77, 1970.

"Air Pollution Control in Texas," Texas Law Review, 47:1086 (June, 1969).

"Air Pollution in the San Francisco Bay Area," Stanford Workshop on Air Pollution, Standford University, Stanford, California, 1970.

"Air Pollution Legislation in Kentucky: Including a Comparative Study of Control Statutes in Pennsylvania, New York, California, Ohio, and Indiana." Kentucky Legislative Research Committee on Air Pollution, Control Systems. Research Report No. 33.

Air Pollution, 1969. Hearings of October 27, 1969, on Problems and Programs Associated with the Control of Air Pollution, U.S. Senate, Committee on Puclic Works, Subcommittee on Air and Water Pollution, 1970.

✓ "Air Pollution, Pre-Emption, Local Problems and the Constitution-- Some Pigeonholes and Hatracks," Arizona Law Review, 10:97 (Summer, 1968).

"Air Pollution Symposium, Air Pollution: Causes, Sources, and Abatement: Local Regulations of Air Pollution; State Regulations of Air Pollution; Interstate Agreements for Air Pollution Control; The Federal Air Pollution Program," Washington University Law Quarterly (Spring, 1968), 205-31.

"Air Quality Act of 1967," 42 U.S.C. Sections 1857-57e Supp V, 1970.

"Air Quality Act of 1967," Journal of the Air Pollution Control Association, 18:2 (February, 1968), 62-71.

"Air Quality Act of 1967 Now a Matter of Fact," *Environmental Science and Technology* 1:11 (November, 1967), 884-86.

"Air Quality Criteria for Sulfur Oxides." National Center for Air Pollution Control, Bureau of Disease Prevention and Environmental Control, U. S. Department of Health, Education, and Welfare. Public Health Service, Pub. No. 1619, 1967.

American Law Institute. "Restatement of Torts (Second)," 1965.

"Arkansas' Air Pollution Control Code as it Affects Municipalities," *Arkansas Municipalities* (September, 1969), 16-17.

Atwood, J. R. "An Economic Analysis of Land Use Conflicts," *Stanford Law Review*, 21 (1969), 293-315.

Ayers, Stephen M. "The Effects of Air Pollution upon Health." Proceedings of the Seminar on Legal Aspects of Air Pollution Rutgers, the State University, May 6, 1967, pp. 17-27.

Ayres, R. U. "Air Pollution in Cities, *Natural Resources Journal*, 9 (1969), 1-22.

Bagge, C. E. "Affluence and Effluents: The Challenge of Environment Quality Control," *Public Utilities Fortnightly*, 79 (April 27, 1967), 58-68.

Ballman, H. C. "The Role to be Played by Local and State Government." Proceedings of the National Conference on Air Pollution, 1963. Public Health Service, Pub. No. 1022, pp. 324-26.

Bastian, R. E. "Air Basin Pollution Control," *Pennsylvania's Health*, 30:2 (Summer, 1969), 10-12.

Baxter, W.F. "The SST: From Watts to Harlem in Two Hours," *Stanford Law Review*, 21 (1968), 1-57.

Bell, A. W. and G. T. Norvell. "Air Pollution Control in Texas," *Texas Law Review*, 47 (1969), pp. 1086-1123.

Bellis, I. H., H. F. Kolsby, and E. L. Wolf. "Legal Approach to Industrial Pollution," *Pennsylvania Bar Association Quarterly*, 40 (October, 1968) 96: *Trial*, 4 (June-July, 1968), 29.

Berger, C. M. "Air Pollution as a Private Nuisance, *Washington and Lee Law Review*, 24 (1967), pp. 314-19.

Bloomfield, B. D. "The Foundry--and Air Pollution Control Legislation," Modern Castings, 52:4 (October, 1967), 3-97.

Borchers, D. J., and L. A. Miller. "Private Lawsuits and Air Pollution Control," ABAJ, 56 (1970), 465-69.

_____. "The Practice of Regional Regulation Under the Clear Air Act," Natural Resources Lawyer, 3 (1970), 59-65.

Brestel, W.C., Jr. "The California Motor Vehicle Pollution Control Law," California Law Review, 50 (1962), 121-30.

Cahn, "Environmentalists Blaze Legal Trail to Preserve Nature" Christian Science Monitor, October 2, 1969, page 3, Col. 1.

Calabresi, G. "Does The Fault System Optimally Control Primary Accident Costs?" Law and Contemporary Problems, 33 (1968), 429-63.

_____. "Fault, Accidents and the Wonderful World of Blum and Kalven," Yale Law Journal, 75 (1965), 216-38.

_____. "Some Thoughts on Risk Distribution and the Law of Torts," Yale Law Journal, 70 (1961), 499-553.

_____. The Cost of Accidents. New Haven: Yale University Press, 1970.

_____. "The Decision for Accidents: An Approach to Nonfault Allocation of Costs," Harvard Law Review, 78 (1968), 713-45.

_____, "Transaction Costs, Resource Allocation and Liability Rules--A Comment," Journal of Law and Economics, 11 (1968), 67-73.

Calvert, S., and W. R. Chalkner, "Federal Role in Emission Criteria and Standards." Proceedings of the Third National Conference on Air Pollution. Public Health Service, Pub. No. 1649, 1967. pp. 476-79.

"Camara (Camara v. Municipal Court of the City and County of San Francisco, 87 Sup. Court 1727) and See (See v. Seattle, 87 Sup. Court 1737): A Constitutional Problem with Effect on Air Pollution Control," Arizona Law Review, 10 (Summer, 1968), 120.

Carter, L. J. "Conservation Law II: Scientists Play a Key Role in Court Suits," Science, 166 (1969), 1601-6.

_____. "Conservation Law II: Seeking a Breakthrough in the Courts," Science, 166 (1969), 1487-91.

Carver, J. A. "Pollution Control and the Federal Power Commission," Natural Resources Law, 1 (January, 1968), 32.

Cassell, E. J. "Are We Ready for Ambient Air Quality Standards?" Journal of the Air Pollution Control Association, 18:12 (December, 1968), 799-802.

_____. "Health Effects of Air Pollution and Their Implications for Control," Law and Contemporary Problems, 33:2 (Spring, 1968).

Chass, R. L., and E. S. Feldman. "Tears for John Doe," Southern California Law Review, 27 (1954), 349-72.

Clary, J. T. "Air and Water Interstate Compacts," Natural Resources Law, 1 (October, 1968), 60.

"Clean Air Amendments of 1970", Public Law No. 91-604, Section 4a, 1970.

Coase, R. H. See Economics Section.

Cohen, J. L. "Interstate Compacts--An Evaluation," Journal of the Air Pollution Control Association, 17:10 (October, 1967), 676-78.

Committee on Pollution, Waste Management and Control National Academy of Sciences--National Research Council, 1966.

Conti, "Conservationists Press Suits to Assert Right to a Clean Environment," Wall Street Journal, March 26, 1970 (Pacific Coast ed.), p. 1, col. 6.

"Controlling Air Pollution from Garden State, New Jersey, Incinerators; Chapter 11 of New Jersey Air Pollution Code Gives State Health Department Power of Enforcement," New Jersey Municipalities (June, 1969), 13-14.

Cowan, C. A. "Air Pollution Control in New Jersey," Rutgers Law Review, 9 (1955), 609-33.

Cross, T. W. "The Provincial Air Pollution Control Program as It Relates to the Municipaliteis," Municipal World (January, 1968), 3-5.

Currie, D. P. "Trail Blazers at Law," Trial, 5:23 (1969), 28.

Dales, J. H. See Economics Section.

Daley, A. C. "Problems in Statewide Uniform Air Quality Enforcement," Journal of the Air Pollution Control Association, 19:2 (February, 1969), 77-80.

Davis, K. C. "Judicial Notice," Columbia Law Review, 55 (1955), 945-84.

Delogu, O. E. "Effluent Charges: A Method of Enforcing Stream Standards," Maine Law Review, 19 (1967), 29-47

_____. "Legal Aspects of Air Pollution Control and Proposed State Legislation for Such Action," Wisconsin Law Review (1969), 884.

DeVany, A. S., R. D. Eckert, C. J. Meyers, D. J. O'Hara, and R. C. Scott. "A Property System for Market Allocation of the Electromagnetic Spectrum: A Legal-Economic-Engineering Study, Stanford Law Review, 21 (1969), 1499-1561.

Digest of Municipal Air Pollution Ordinances. Washington, D.C.: Public Health Service, U.S. Department of Health, Education, and Welfare, 1962.

Dingell, J. D. "A Congressional View of the Federal Role in Air Pollution Control." Proceedings of the Third National Conference on Air Pollution. Public Health Service, Pub. No. 1649, 1967, pp. 516-18

Drowley, W. B. "The Problem of Establishing Guides and Standards for Air Pollution Control." Pollution and Our Environment. Background Paper C22-2. Montreal: Canadian Council of Resource Ministers, 1967, pp. 1-15.

Edelman, S. "Air Pollution Abatement Procedures Under the Clean Air Act," Arizona Law Review, 10 (1968), 30-36.

_____. "Air Pollution Control Legislation." Air Pollution. A. C. Stern, ed. Second edition, Volume 3. New York: Academic Press, 1968, pp. 553-59.

_____. "Concluding Remarks." Proceedings of the National Conference on Air Pollution. Public Health Service, Pub. No. 1022, 1963, pp. 337-38.

_____. "Legal Problems of Interjurisdictional Air Pollution," Journal of the Air Pollution Control Association, 13 (July, 1963), 310-13.

_____. "The Law of Federal Air Pollution Control," Journal of the Air Pollution Control Association, 16:10 (October, 1966), 523-25.

Edwards, M. N. "Legislative Approach to Air and Water Quality," Natural Resources Law, 1 (January, 1968), 58.

Esposito, J. C. See Political Science Section.

"Expanding Scope of Air Pollution Abatement," West Virginia Law Review, 70 (February, 1968), 195.

Finch, R. H. "Air Pollution Control--A Crucial Year," Minerals Process, 10:5 (May, 1969), 9-10.

Fitzpatrick, J. V. "Regional Approach to Community Air Pollution Control," American Journal of Public Health, 55:90 (June, 1965), 5-10.

_____. "The Role to be Played by the Federal Government." Proceedings of the National Conference on Air Pollution. Public Health Service, Pub. No. 1022, 1963, pp. 329-31.

Fromson, J., "A History of Federal Air Pollution Control," Ohio State Law Review 30 (1969), 516-536.

Fuller, L. "Adjudication and the Rule of Law," Proceedings of the American Society of International Law (1960), 1-8.

Gaulding, C. L. "Definitions of Air Quality Control Regions Approach and Experience to Date," Journal of the Air Pollution Control Association, 18:9 (September, 1968), 591-95.

Gerhardt, P. H. "Incentives to Air Pollution Control," Law and Contemporary Problems, 33 (1968), 358-68.

German, R. H. "Regulation of Smoke and Air Pollution in Pennsylvania," University of Pittsburgh Law Reveiw, 10 (1949), 493-510.

Goldman, M. I., ed. Pollution: The Mess Around Us. Englewood Cliffs, N.J.: Prentice-Hall, 1967.

Goldner, L. "Air Pollution Control in the Metropolitan Boston Area: A Case Study in Public Policy Formation." The Economics of Air Pollution. H. Wolozin, ed. New York: W. W. Norton and Co., 1966.

Goodman, F. "Air Pollution Ordinances," Baylor Law Review, 8 (1956), 249-56.

"Government and Industry Work Out Model Rule," Modern Manufacturing 2:7 (July, 1969), 168-70.

"Government Programs to Encourage Private Investment in Low Income Housing," Harvard Law Review, 81 (1968), 1295-1324.

Green, L. C. "State Control of Interstate Air Pollution," Law and Contemporary Problems, 33:1 (Spring, 1968), 369-98.

Gruber, C. W. "The Role to be Played by Local and State Government." Proceedings of the National Conference on Air Pollution. Public Health Service, Pub. No. 1022, 1963, pp. 322-23.

Grundy, R. "Rationale for Air Quality Criteria," Environmental Science and Technology, 2:10 (October, 1968), 742-49.

Haar, C. M. Land-Use Planning. Boston: Little, Brown and Co., 1959.

Hagevik, G. "Legislating for Air Quality Management: Reducing Theory to Practice," Law and Contemporary Problems, 33:2 (Spring, 1968), 369-98.

Hairden, M.C., and W. M. Anderson. "Legal Aspects of Air Pollution," Virginia Journal of Science, 18:2 (1967), 77-79.

Hamel, Q. S. "Air Pollution--Automobile Smog: A Proposed Remedy," DePaul Law Review, 14 (1965), 436-44.

Hamill, L. "The Process of Making Good Decisions About the Use of the Environment of Man," Natural Resources Journal 8 (1968), 279-301.

Hanks, E., and J. Hanks. "An Environmental Bill of Rights: The Citizen Suit and the National Environmental Policy Act of 1969," Rutgers Law Review, 24 (1970), 230-72.

Hardin, G. "The Tragedy of the Commons," Science, 162 (1968), 1243-48.

Harper, F., and F. James. The Law of Torts. Boston: Little, Brown and Co., 1956.

Havihurst, C. C. "Foreword," Law and Contemporary Problems, 33 (1968), 195-96.

Haydel, D. "Regional Control of Air and Water Pollution in the San Francisco Bay Area," California Law Review (August, 1967), 702-27.

Herfindahl, O. C., and A. V. Kneese. See Economics Section.

High, M. D., and W. H. Megonnell. "Pollution Control, Federal Leadership," Mechanical Engineering, 91:2 (February, 1969), 20-23.

Hines, N. W. "Nor Any Drop to Drink: Public Regulation of Water Quality," Iowa Law Review, 52 (1969), 186-235.

"History of Federal Air Pollution Control," Ohio State Law Journal, 30 (Summer, 1969), 516.

Hochheiser, Seymour. "Air Pollution Detection and Control," Virginia Journal of Science, 182:2 (1967), 76-79.

Hurst, J. W. Law and Social Process in United States History (Ann-Arbor: University of Michigan Law School, 1960).

"International Air Pollution--U.S. and Canada--A Joint Approach," Arizona Law Review, 10:138 (Summer, 1968).

Jackson, W. E., and H. C. Wohlers. "You Need an Air Pollution Inventory . . . to Set Meaningful Air Purity Standards in Any Affected Areas: Here's the Way It Was Done in the Delaware Valley," American City (October, 1967), 119-20.

Jaffe, L. J. "The Citizen as Litigant in Public Actions: The Non-Hohfeldian or Ideological Plaintiff." University of Pennsylvania Law Review 116 (1968), 1033-47.

_____. "Standing to Secure Judicial Review: Public Actions," Harvard Law Review, 74 (1961), 1265-1314.

_____. "Standing to Sue in Conservation Suits." in Law and the Environment, M. Baldwin and J. Page, eds. Walker and Co., 1970.

Johnson, E. P. "The Expanding Scope of Air Pollution Abatement," West Virginia Law Review 70 (1968), 195-203.

Jordon, F. J. E. "Recent Developments in International Environmental Pollution Control," McGill Law Journal, 15 (June, 1969), 279.

Juergensmeyer, J. C. "Control of Air Pollution Through the Assertion of Private Rights," Duke Law Journal (1967), 1126.

_____, and A. L. Morse. "Air Pollution Control in Indiana in 1968: A Comment," Valparaiso University Law Review, 2 (1968), 296-314.

Katz, M. "The Function of Tort Liability in Technology Assessment," University of Cincinnati Law Review, 38 (1969), 587-662.

Keagy, D. M. "The Clean Air Act and Its Effects." Paper presented at the Second Annual Conference, Pacific Northwest International Section of the Air Pollution Control Association, Portland, Oregon, November, 1964.

Kennan, G. F. "To Prevent a World Wasteland," Foreign Affairs, 48 (1970), 401-13.

Kennedy, H. W. "Legal Aspects of Air Pollution Control," Public Health Reports, 79 (August, 1964), 689-98.

_____. "Legal Aspects of Community Air Quality Standards." Unpublished manuscript, 1962.

_____. "Legal Aspects of Human Exposure to Atmospheric Pollutants." Unpublished manuscript, 1961.

_____. "Legal Support for Los Angeles County's Strict Air Pollution Control Program." Report to the Los Angeles County Board of Supervisors, September, 1957.

_____. "Legislation and Environmental Health," Archives of Environmental Health, 12:1 (January, 1966), 129-35.

_____. "Levels of Responsibility for the Administration of Air Pollution Control Programs." Proceedings of the National Conference on Air Pollution. Public Health Service, Pub. No. 645, 1959.

_____. "Some Legal and Public Policy Problems in Air Conservation." Unpublished manuscript, 1962.

_____. "The Formulation and Adoption of Reasonable Rules and Regulations." Unpublished manuscript, 1962.

_____. "The Legal Aspects of Air Pollution with Particular Reference to the County of Los Angeles," Southern California Law Review 27 (1954), 373-98.

_____. "The Mechanics of Legislative and Regulatory Action." Proceedings of the National Conference on Air Pollution. Public Health Service, Pub. No. 1022, 1963, pp. 306-14.

_____, and A. O. Porter. "Air Pollution: Its Control and Abatement," Vanderbilt Law Review 8 (1955), 854-77.

_____, and Martin E. Weeks. "Control of Automobile Emissions-- California Experience and the Federal Legislation," Law and Contemporary Problems, 33:2 (Spring, 1968), 297-314.

Kerner, O. "A State View of the Federal Role." Proceedings of the Third National Conference on Air Pollution. Publich Health Service, Pub. No. 1048, 1967, pp. 508-11.

Kimball, G. J. "The Application of Res Ipsa Loquitur in Suits Against Multiple Defendants," Albany Law Review, 34 (1969), 106-21.

Kneese, A. V. "How Much Is Air Pollution Costing Us in the United States?" Proceedings of the Third National Conference on Air Pollution. Public Health Service, Pub. No. 1649, 1967, pp. 529-37.

_____. "Statement." Hearings Before a Special Subcommittee on Air and Water Pollution, Senate Committee on Public Works. 89th Congress, 1st. Session, 1965.

Kovel, A. "A Case for Civil Penalties: Air Pollution Control," Journal of Urban Law, 46 (1968), 156.

Krier, J. E. "Environmental Litigation and the Burden of Proof." Law and the Environment. M. Baldwin and J. Page, Walker and Co., 1970.

_____. "Environmental Watchdogs: Some Lessons From a Study Council," Stanford Law Review, 23 (1971).

Krutilla, J. V. See Economics Section.

Kushner, Adele. "Can we Control our Dirty Air?" Atlanta Economic Review, 17 (September, 1967), 9-13.

Larsen, R. I. "Determining Reduced-Emission Gas Needed to Achieve Air Quality Goals . . . A Hypothetical Case," <u>Journal of the Air Pollution Control Association</u>, 17 (December, 1967), 823-29.

Leduc, E. C. "Constructing Air Pollution Enforcement Program Boundaries." Paper presented at the sixty-second annual meeting of the Air Pollution Control Association, New York, June 22-26, 1969.

<u>Legal Road to Cleaner Air</u>. Mid-Atlantic States Section, Air Pollution Control Association, January, 1969.

Lewis, H. R. "A Time for Imagination." Proceedings of the Third National Conference on Air Pollution. Public Health Service, Pub. No. 1649, 1967, pp. 521-24.

Lieber, H. See Political Science Section.

"Local Regulation of Air Pollution," <u>Washington University Law Quarterly</u>, (1968), 232-48.

MacKenzie, V. G. "Approaches to Air Quality Regulations." Proceedings of the fifty-fourth Air Pollution Control Association Conference. New York, June, 1961.

_____. "The Clean Air Act and Its Implications," <u>American Journal of Public Health</u>, 55 (June, 1964), 901-4.

_____. "The Impact of the Clean Air Act on State and Local Control Programs." Paper presented at the sixth Annual Sanitation Engineering Conference, Pittsburgh, January 9, 1955.

_____. "The New Federal Clean Air Act," <u>Journal of the Air Pollution Control Association</u>, 14 (September, 1964), 385-87.

Maga, J. A. "Air Quality Criteria and Standards." Proceedings of the Third National Conference on Air Pollution. Public Health Service, Pub. No. 1649, 1967, pp. 469-71.

_____. "Air Resource Management in the San Francisco Bay Area." Berkeley: Institute of Governmental Studies, University of California at Berkeley, 1965.

_____. "Considerations in Establishing Air Quality Standards" <u>Journal of Occupations Medicine</u>, 10:8 (August, 1968), 408-13.

_____. "Motor Vehicle Pollution in California," <u>Journal of the Air Pollution Control Association</u>, 17:7 (July, 1967), 435-38.

Mandelker, D. R. "Quality Standards for the Control of Pollution."
Development of Air Quality Standards. A. Atkisson and R. Gaines,
eds., Charles E. Merrill Pub. Co., 1970.

_____. Managing Our Urban Environment: Cases, Texts, and
Problems. New York: Bobbs-Merrill, 1966.

Manual for the Development of State Recommendations for the Air
Quality Control Regions, United States. Washington, D.C.:
National Air Pollution Control Administration, January, 1970.

Martin, R., and L. Symington. "A Guide to the Air Quality Act of
1967," Law and Contemporary Problems, 33:2 (Spring, 1968),
239-74.

McCarthy, R. D. "Recent Legal Developments in Environmental
Defense," Buffalo Law Review, 19 (1970), 195-204.

McCauley, B. T., and D. J. Morgan. "Wyoming Air Quality Act,"
Land and Water Law Review, 4 (1969), 159-84.

McKee, H. C. "Federal Abatement of Major Intrastate Air Pollution
Sources." Proceedings of the Third National Conference on
Air Pollution. Public Health Service, Pub. No. 1649, 1967, pp.
487-90.

McKee, J. W. "Trends in the Control of Photochemical Smog in the
Los Angeles Basin," Journal of Environmental Science, 11:3
(June, 1968), 34-41.

Mees, Q. M. "Air Pollution: The Silent Assault," Arizona Review,
(June-July, 1969), 1-6.

Megonnell, W. H. "Federal Air Pollution Prevention and Abatement
Responsibilities and Operations." Paper presented at the fifty-
ninth annual meeting of the Air Pollution Control Association,
San Francisco, June 20-24, 1966.

_____. "Impact of Pollution Abatement Regulations on Industry and
Municipalities." Paper presented at the Summer Conference on
the Demands of Pollution Control Legislation, Fairleigh Dickinson
University, Madison, New Jersey, August 26, 1966. Washington,
D.C.: Division of Air Pollution, Bureau of State Services, Depart-
ment of Health, Education, and Welfare, 1966.

Michelman, F. I. "Property, Utility, and Fairness: Comments on the
Ethical Foundations of 'Just Compensation' Law," Harvard Law
Review 80 (1967) 1165-1285.

Middleton, J. T. "Summary of the Air Quality Act of 1967," Arizona Law Review, 10 (1968), 25-29.

Miller, L. A., and D. J. Borchers. "Air Pollution Control Through Legal Actions: A Breath of Fresh Air." Paper presented at the sixty-second annual meeting of the Air Pollution Control Association, New York. June 22-26, 1969.

Mills, E. S. See Economics Section.

Mix, D. D. "The Misdemeanor Approach to Pollution Control," Arizona Law Review, 10 (1968), 90-96.

"Model Interstate Compact for the Control of Air Pollution," Harvard Journal of Legislation, 4 (June, 1967), 369-98.

Moody, J. E. "Air Quality Improvement--A Look Ahead," Natural Resource Law, 2:7 (January, 1969).

Moran, G. R. "The Air Pollution Control Act and Its Administration [New Jersey]," Rutgers Law Review, 9 (1955), 640-81.

Murphy, E. F. "A Law for Life," Wisconsin Law Review, 773 (1969), 87.

_____. "Air Pollution: Bibliography of Congressional Material," Law Library Journal, 62 (February-May, 1969), 84-89.

Muskie, E. S. "Environmental Jurisdiction in the Congress and the Executive," Maine Law Review, 22 (1970), 171-87.

_____. "Role of the Federal Government in Air Pollution Control," Arizona Law Review, 10 (1968), 17-24.

Neustadter, G. "The Role of the Judiciary in the Confrontation with the Problems of Environmental Quality," UCLA Law Review, 17 (1970), 1070-1100.

"New Air Quality Control Regions Proposed," Municipal Attorney (May, 1969), 72-73.

Note. "Air Pollution Control in Minnesota," Minnesota Law Review, 54 (1970), 953-977.

_____. "Water Quality Standards in Private Nuisance Actions," Yale Law Review, 79 (1969), 102-110.

_____. (a) "Air Pollution: Causes, Sources and Abatement," Washington University Law Quarterly, (1968), 205-231.

_____. (b) "The Air Quality Act of 1967," Iowa Law Review, 54 (1968), 115-140.

_____. (c) "The Federal Air Pollution Program," Washington University Law Quarterly (1968), 283-324.

_____. (d) "State Regulation of Air Pollution," Washington University Law Quarterly (1968), 249-259.

_____. (e) "Local Regulation of Air Pollution," Washington University Law Quarterly (1968), 232-248.

_____. (f) "Government Programs to Encourage Private Investment in Low Income Housing," Harvard Law Review, 81(1968), 1295-1324.

_____. (g) "A Trend Toward Coalescence of Trespass and Nuisance: Remedy for Invasion of Particulates," Washington University Law Quarterly (1961), 62-73.

_____. (h) "Smog--Can Legislation Clear the Air," Stanford Law Review, 1 (1949), 452-462.

"1968 State Legislation for Pollution Control," Journal of the Air Pollution Control Association, 19:2 (February, 1969), 81-90.

O'Fallon, J. E. "Deficiencies in the Air Quality Act of 1967," Law and Contemporary Problems, 33:2 (Spring, 1968), 275-96.

Polatsek, D. G. "Legal Note, Air Pollution Control," Public Health Report, 81 (October, 1966), 884.

_____. "Legal Note, Air Pollution Control," Public Health Report, 81 (May, 1966), 435-36.

Pollack, L. W. "Legal Boundaries of Air Pollution Control--State and Local Legislative Purpose and Techniques," Law and Contemporary Problems, 33:2 (Spring, 1968), 331-57.

_____. "Legal Techniques of Air Pollution Control--The 1966 New York City Legislation," New York Law Journal, Parts 1-5 (October 17-21, 1966).

Porter, W. C. "The Role of Private Nuisance Law in the Control of Air Pollution," Arizona Law Review, 10 (1968), 107-19.

Prindle, R. A. "Air Quality Criteria and Standards." Proceedings of the Third National Conference on Air Pollution. Public Health Service, Pub. No. 1649, 1967, pp. 465-68.

Prosser, W. L. "Law of Torts." West. (Third ed.), 1964.

Purdom, P. W. "The Role to Be Played by Local and State Government." Proceedings of the National Conference on Air Pollution. Public Health Service, Pub. No. 1022, 1963, pp. 317-21.

Rayher, W. A., and J. T. Middleton. "The Case for Clean Air," Mill and Factory, 80 (April, 1967), 41-56.

"Regional Approach to Air Pollution Abatement Gains Momentum," Environment Science and Technology, 3 (May, 1969), 431-34.

"Regional Attack on Air Pollution," Wharton Quarterly, 3 (Spring, 1969), 31-39.

"Regional Control of Air and Water Pollution in the San Francisco Bay Area," California Law Review, 55 (August, 1967), 702.

Reich, C. "The New Property," Yale Law Journal (April, 1964), 733-87.

Reitze, A. W., Jr. "Pollution Control: Why Has it Failed?" ABAJ, 55 (1969), 923-27.

_____. "Role of the 'Region' in Air Pollution Control," Case Western Reserve Law Review, 20 (August, 1969), 809.

Rempe, G. A. "International Air Pollution--United States and Canada--A Joint Approach," Arizona Law Review, 10 (1968), 138-47.

Rheingold, P. D. "Civil Cause of Action for Lung Damage Due to Urban Atmosphere," Brooklyn Law Review, 33 (1968), 17.

Richards, J. E. "Further Comment on the Clean Air Act of 1968," Smokeless Air, 149 (Spring, 1966), 197-201.

Ricksen, R. H. "Legislative Limitation on Air Pollution Enforcement," Hastings Law Journal, (1958), 191-98.

Roberts, K. A. "The Role to Be Played by the Federal Government." Proceedings of the National Conference on Air Pollution. Public Health Service, Pub. No. 1022, 1963, pp. 327-28.

Roberts, M. J. "River Basin Authorities: A National Solution to Water Pollution," Harvard Law Review, 83 (1970), 1527-56.

Roemer, R., J. Frink, and C. Kramer. "Environmental Health Services: Multiplicity of Jurisdictions and Comprehensive Environmental Management," Milbank Memorial Fund Quarterly, 49:3 (1971).

Rogers, S. M. "A Review and Appraisal of Air Pollution Legislation in the U. S.," Journal of the Air Pollution Control Association (February, 1958), 308-15.

_____, and S. Edelman. A Digest of State Air Pollution Laws [Since 1961]. Public Health Service, Pub. No. 711.

"Role of Private Nuisance Law in the Control of Air Pollution," Arizona Law Review, 10 (Summer, 1968), 107.

Rossano, A. T. "Federal Abatement of Major Intrastate Air Pollutoin Sources." Proceedings of the Third National Conference on Air Pollution. Public Health Service, Pub. No. 1649, 1967, pp. 480-86.

Ruff, L. E. "The Economic Common Sense Of Pollution," Public Interest (Spring, 1970), 69-85.

"Rules and Regulations of the Air Pollution Control District of Los Angeles County" Air Pollution Control Engineering Manual. Public Health Service, Pub. No. 999-AP-40, pp. 831-57.

Sax, J. L. "Public Rights in Public Resources: The Citizen's Role in Conservation and Development." Contemporary Developments in Water Law. C. Johnson and S. Lewis, eds. Center for Research in Water Resources, 1970.

_____. "The Public Trust Doctrine in Natural Resource Law: Effective Judicial Intervention," Michigan Law Review, 68 (January, 1970), 471-566.

Schmitz, T. M. "Pollution, Law, Science and Damage Awards," Cleveland State Law Review, 18 (September, 1969), 467.

Schoonover, B. L. "Public Relations, Law, Environmental Pollution," Cleveland State Law Review, 18 (September, 1969).

Schrenk, H. "Air Pollution in Donora, Pennsylvania, Epidemiology of the Unusual Smog Episode of October, 1948." Public Health Service, Bulletin No. 306, 1949.

Schrunk, T. D. "A Local View of the Federal Role." Proceedings of the Third National Conference on Air Pollution. Public Health Service, Pub. No. 1649, 1967, pp. 503-7.

Schueneman, J. J. "Air Pollution Control Administration." Air Pollution. Volume 3. A. C. Stern, ed. Second edition. Academic Press, 1968, pp. 719-96.

Schwartz, S. "Letter to the Editor," Los Angeles Times, December 20, 1969, Part 2, p. 4.

Seamans, F. C. "Tort Liability for Pollution of Air and Water," Natural Resources Lawyer, 3 (1970), 146-53.

Shapiro, F. C. "Our Far-Flung Correspondents, Environmental Disruption," New Yorker (May 23, 1970), 93-105.

Simon, A. "Battle of Beaufort," New Republic, 162 (1970), 11-15.

Smith, S., and Griswold. "Case Study: Los Angeles Battles the Air Trap," Public Management, 49 (April, 1967), 84-88.

"Smog--Can Legislation Clean the Air?" Stanford Law Review, 1 (1949), 452-62.

Source Materials of Air Pollution Control Laws. Public Air Quality Committee, Manufacturing Chemists Association, Washington D.C.: June, 1967.

Staff of the Senate Commission on Public Works. "A Study of Pollution--Air." 88th Congress, 1st Session, 1963.

Stalker, W. W., and C. B. Robinson. "A Method for Using Air Pollution Measurements and Public Opinion to Establish Ambient Air Quality Standards," Journal of the Air Pollution Control Association, 17:4 (March, 1967), 142-44.

"State Air Pollution Control Legislation," Boston College Industrial and Commercial Law Review, 9 (Spring, 1968), 712.

State Legislative and Constitutional Action on Urban Problems in 1967. Washington, D.C.: Advisory Commission on Intergovernmental Relations, April, 1968. CFSTI PB 178982.

"State Regulation of Air Pollution," Washington University Law Quarterly, 249-59.

"Status of Environmental Quality Legislative Measures," Environmental

Science and Technology, 2 (October, 1968), 755-58.

Steinberg, L. W. "Rights Under California Law of the Individual Injured by Air Pollution," Southern California Law Review, 27 (1954), 405-14.

Stern, A. C. "Implications of the Air Quality Act of 1967," New York Academy of Sciences Transactions, 30:6 (April, 1968), 759-65.

Stites, J. G. "Federal Support for Industrial Air Pollution Control Methods." Proceedings of the Third National Conference on Air Pollution. Public Health Service, Pub. No. 1649, 1967, pp. 499-502.

Stumph, T. L. and R. L. Duprey. "Trends in Air Pollution Control Regulations." Paper presented at the sixty-second annual meeting, Air Pollution Control Association, New York, June 22-26, 1969.

"Summary of Air Pollution Actions in 1966 State Legislatures," Journal of the Air Pollution Control Association, 17 (February, 1967), 115-18.

Sussman, V. H. "The Utilization of Ambient Air Quality Criteria and Standards," Journal of the Air Pollution Control Association, 19:2 (February, 1969), 73-96.

Thayer, H. E. "An Industrial View of the Federal Role." Proceedings of the Third National Conference on Air Pollution. Public Health Service, Pub. No. 1649, 1967. pp. 512-15.

"The Air Quality Act of 1967," Iowa Law Review, 54 (1968), 115-40.

"The Federal Air Pollution Program," Washington University Law Quarterly, 283-324.

"Three Major Goals for Air Purity," American City, 82 (July, 1967), 16.

"Towards a Clean Air Policy." Proceedings of the International Clean Air Congress, 1966. Paper No. VI/16, pp. 203-5.

Trial, 5:10 (August-September, 1969), "Can Law Reclaim Man's Environment?" Trial Enters a Think Tank (McLanahan, Phoenix, Searcy, Miller); V. J. Yannacane, Jr. "A Lawyer Answers the Technocrats." E. S. Muskie, "Public Commitment First." W. L. Guy, "Needed: U. S. Planning Agency." R. E. Train, "Crimes Against the Environment." W. H. Ferry, "For Effective Control: Revise the Constitution?" D. P. Currie, "Trail Blazer-at-Law." J. S. Sax, "Slumlordism: Another Pollutant-- A New Tort." B. S. Cohen, "Legal Defense of Environmental Rights."

Tukey, J. W. "Status Report." Proceedings of the Third National Conference on Air Pollution. Public Health Service, Pub. No. 1649, 1967, pp. 462-64.

Tyler, R. "Methods for State Level Enforcement of Air and Water Pollution Laws," Texas Bar Journal, 31 (November, 1968), 905.

Verleger, P. K., and J. M. Crowley. "Air Pollution, Water Pollution, Industrial Cooperation and the Antitrust Laws," Land and Water Law Review, 4 (1969), 475.

_____. "Pollution: Regulation and the Antitrust Laws," Natural Resources Law, 2 (May, 1969), 131.

Walderman, H. "Legal Note: Air Pollution Control," Public Health Report, 83: 2 (February, 1968), 118.

Walker, M. S. "Enforcement of Performance Requirements With Injunctive Procedure," Arizona Law Review, 10 (1968), 81-89.

Warren, David G. "State and Local Regulation of Air Pollution," Popular Government, 33 (February, 1967), 1-10.

"Waste Management and Control." Committee on Pollution, National Academy of Sciences-National Research Council, 1966.

"Water Quality Standards in Private Nuisance Actions," Yale Law Journal, 79 (1969), 102-10.

Welsh, G. E. "Federal Air Quality Act of 1967," Natural Resources Lawyer, 3 (1970), 52-58.

Wohlers, H. C., P. W. Purdom, et al. "Can Air Pollution Be Controlled by Legislation?" Scientia, 104: 681-82, Series 8 (January-February, 1969), 58-64.

Wolf, E. L. "Legal Aspects of Air Pollution," Trial Law Quarterly, 5 (Winter, 1968), 22.

Wolozin, H. See Economics Section.

Working Committee on Economic Incentives. "Cost Sharing with Industry?" Washington, D.C.: Federal Coordinating Committee in the Economic Impact of Pollution Abatements, 1967.

Wright, G. A. "The Cost-Internalization Case for Class Action," Stanford Law Review 21 (1969), 383-419.

Yaffe, C. D. "Progress in State and Local Air Pollution Control Under the Clean Air Act." Paper presented at the American Industrial Hygiene Conference, Denver, May 14, 1969.

Yannacone, V. J. "A Lawyer Answers the Technocrats," Trial, 5 (August-September, 1969), 14-15.

Zimney, T. L. "The Peril of Air Pollution in North Dakota," North Dakota Law Review, 46 (1970), 217-38.

PLANNING

"Air Conservation." Washington, D.C.: Air Conservation Commission, American Association for the Advancement of Science, 1965. Publication No. 80.

"Air Pollution: A Growing Urban Problem." Chicago: American Society of Planning Officials, 1955. Report No. 79.

"Air Pollution and the Siting of Industry." Proceedings of the Fourth New Zealand Geography Conference, 1965, pp. 161-65.

"Air Pollution: A Non-Technical Bibliography." Monticello, Ill.: Council of Planning Librarians, 1969.

"Air Pollution Control." Chicago: American Society of Planning Officials, 1950. Report No. 212.

"Air Zoning: An Application of Air Resource Management." Chicago: American Society of Planning Officials, 1966. Report No. 212.

Arnold, G. and E. Edgerly, Jr. "Urban Development in Air Pollution Basins: An Appeal to the Planners for Help," Journal of the Air Pollution Control Association, 17 (April, 1967), 235-37.

Atkisson, Arthur A. "Air Contaminants as Factors in Industrial Land Use Planning and Zoning." Paper presented before AIP Northern California Section, San Jose, 1956.

Boffey, P. M. "Smog: Los Angeles Running Hard, Standing Still," 161 (1968), 990-92.

Branch, M. C. "Comprehensive Urban Planning: A Selective Annotative Bibliography with Related Materials." Sage Publications, 1970.

Braybrooke, D., and C. Lindblom. A Strategy of Decision. New York: Free Press, 1963.

Breivogel, Milton, et al. "Air Pollution-Potential Advisory Service for Industrial Zoning Cases." LAPCD, Air Pollution Control Association, Paper No. 60-39, 1960.

"Building the Good City." German Federation for Housing and Regional Planning. Chapter 5, Pub. No. 76, 1968.

Bunyard, F. L., and J. D. Williams. "Interstate Air Pollution Study-- St. Louis Area Air Pollutant Emissions Related to Actual Land Use," Journal of the Air Pollution Control Association, 17 (April, 1967), 215-19.

Burton, E. S., and William Sanjour. "A Simulation Approach to Air Pollution Abatement Program Planning," Socio-Economic Planning Sciences, 4 (1970), 147-59.

Caldwell, L. K. "Environment: A Focus for Public Policy?" Public Administration Review, 23 (1963), 132-39.

Camblin, G. "Environmental Problems in Town Planning," Public Health Inspector, 76:13 (October, 1968), 553-62, 585.

Chermayeff, Serge. "Design as Catalyst," Socio-Economic Planning Sciences, 1 (1967), 63-69.

Crisis: Air. Eugene, Oregon: Central Lane Planning Council, August, 1968.

"Do Your City Planners Know About Air Pollution?" American City, 82 (April, 1967), 91-94.

Faltermayer, E. K. Redoing America, A Nationwide Report on How to Make Our Suburbs and Cities Livable. New York: Harper and Row, 1968.

Fisher, J. L. "Environment Quality and Urban Living." Planning 1967. Selected papers from the American Society of Planning Officials Conference, 1967.

Fitzpatrick, J. V., and A. N. Heller. "Dynamic Air Resource Management Program, City of Chicago," Journal of the Air Pollution Control Association, 15:7 (1965).

Frenkiel, F. "Atmospheric Pollution and Zoning in an Urban Area." Paper presented at the forty-eighth annual meeting of the Air Pollution Control Association, 1955.

Frieden, Bernard J., and R. Morris. Urban Planning and Social Policy. New York: Basic Books, 1968.

Garnett, Alice. "Some Climatological Problems in Urban Geography with Reference to Air Pollution," Institute of British Geographers, 42 (1967), 21-44.

Gartner, Irvin. "Levels of Program Development of Regional, State and Local Air Pollution Control Agencies." Proceedings of the Air Pollution Control Association, 1968.

George, E. R., Julien A. Verssen, and R. L. Choss. "A Growing Pollution Source." Paper presented at the sixty-second annual meeting of the Air Pollution Control Association, 1969.

Goldner, L. "Air Pollution Control Should Be a Built-in Element of Community Development," Journal of Housing, 8 (September, 1964), 414-18.

_____. Air Pollution Effects and Planning. Washington, D.C.: U. S. Department of Health, Education, and Welfare, Public Health Service, 1963.

"Guidelines for the Development of Air Quality Standards and Implementation Plans." Washington, D.C.: Public Health Service, National Air Pollution Control Administration, Department of Health, Education and Welfare, 1969.

Herzog, H. W., Jr. "The Air Diffusion Model as an Urban Planning Tool," Socio-Economic Planning Sciences, 3:4 (December, 1969), 329-49.

Hoch, Irving. "Trade-offs Involving City Size, Density and Building Type." Washington, D.C.: Institute of Public Administration, 1969.

Holland, W. D. "Industrial Zoning as a Means of Controlling Area Source Pollution," Journal of the Air Pollution Control Association, 10:4 (April, 1960), 147-55.

Hufschmidt, M. "Environmental Planning," American Behavioral Scientist (1969), 6-8.

_____. "The Methodology of River Basin Planning." Readings in Resource Management and Conservation. Chicago: University of Chicago Press, 1965, pp. 558-70.

Ingrum, W. T. "Place of Performance Standards in Planning and Zoning Regulations," Journal of the Air Pollution Control Association, 12:63 (1962).

"Institutions for Effective Management of the Environment," Part I; "Environmental Problems in South Florida," Part II, Washington, D. C.: Report of the Environmental Study Group, National Academy of Science, 1970.

Isard, W. "The Linkage of Socio-Economic and Ecological Systems," The Regional Science Association Papers, 21:79, 1968.

Kates, R. "Comprehensive Environmental Planning," Regional Planning: Challenge and Prospects. M. Hufschmidt, ed. New York: Praeger, 1969.

Katz, M. "City Planning, Industrial Plan Location and Air Pollution," Air Pollution Handbook. Parl McGill et al., eds. New York: McGraw-Hill, 1956.

Leavitt, J. M. "Meteorological Considerations in Air Quality Planning," Journal of the Air Pollution Control Association, 10:6 (June, 1960), 246-50.

_____. "Winds as a Factor in City Planning," Landscape, (Autumn, 1957), 16.

Levin, M. Community and Planning: Issues in Public Policy. New York: Praeger, 1969.

Los Angeles City Planning Department. "Extracts from Papers Presented at the Systems Workshop on Vehicle Contaminants Control at the University of Southern California; Ideas and Topics of Particular Interest to Planners," January 24-26, 1968.

Lowry, W. P. "The Climate of Cities," Scientific American, 217 (1967) 15-23.

Maga, J. A. "Air Resource Management in the San Francisco Bay Area." Berkeley: Institute of Governmental Studies, 1965.

"Managing the Air Resource in Northeastern Illinois." Chicago: Northeastern Illinois Planning Commission, 1967. Technical Report No. 6.

Marriott, J. "Public Health Inspector's Role in Planning and Building Control," Public Health Inspector (London), 74:10 (July, 1966), 423-31.

May, R. "The Proper Role of Planning and Zoning in Air Pollution Control." Proceedings of the National Conference on Air

Pollution, Washington, D.C., November 18-20, 1959. Public Health Service, Pub. No. 654. pp. 412-15.

McGrath, D. C., "Planning and Zoning--Can They Be Made to Work for Clean Air? Proceedings of the National Conference on Air Pollution, 1966, pp. 554-64.

Mukherji, A. "Abatement of Atmospheric Pollution by Urban Planning," Traffic Quarterly, 22:3 (July, 1968), 433-50.

Munn, R. E. "The Application of Air Pollution Climatology to Town Planning," International Journal of Air Pollution, 1 (January, 1959), 276-87.

Muskie, E. S. "Computers, Environmental Planning, and the Quality of Life." Proceedings of the IBM Scientific Computing Symposium on Water and Air Resource Management, Yorktown, New York, October 23-25, 1967.

O'Harrow, D. "Performance Standards in Industrial Zoning." Chicago: American Society of Planning Officials, 1957.

Olgyay, V. Design with Climate: Bioclimatic Approach to Architectural Regionalism. Princeton, N. J.: Princeton University Press, 1963.

Pelle, W. J. "Integrating City and Regional Planning with Air Pollution Control--Focus on Northeastern Illinois Metropolitan Area." Paper presented at the sixtieth annual meeting of the Air Pollution Control Association, Paper No. 67-27, 1967.

Perkins, W. A. "Some Effects of the City Structure on the Transport of Airborne Material in Urban Areas." Symposium: Air Over Cities. Washington, D.C.: Public Health Service, Technical Report A62-5, pp. 197-205.

Perloff, H. S. The Quality of the Urban Environment: Essays on "New Resources" in an Urban Age. Baltimore: Resources for the Future, Johns Hopkins University Press, 1969.

"Planning Aspects of Air Pollution Control: Annotated Bibliography." Washington, D.C.: U. S. Public Health Service, Department of Health, Education and Welfare.

"Plant Sites Report, Air and Water," Chemical Weekly, 103:4 (October 5, 1968), 94-98.

Rihm, A., Jr. "Air Pollution and Urban Planning," Health News (Albany: New York State Department of Health, 1962), 15-119.

Rydell, C. P., and Douglas Collins. "Air Pollution and Optimal Urban Form." New York: Hunter College Urban Research Center, City University of New York, June, 1967. Mimeographed.

_____, and G. Schwartz. "Air Pollution and Urban Form: A Review of Current Literature," American Institute of Planners Journal, 34 (March, 1968), 115-20.

Savas, E. S. "Computers in Urban Air Pollution Control Systems," Socio-Economic Planning Sciences, 1 (December, 1967), 157-83.

Schulze, E. E. "Performance Standards in Zoning," Journal of the Air Pollution Control Association 10 (1960), 156.

Seymour, W. N. "Cleaning up Our City Air--Proposals for Combating Air Pollution Through Affirmative Government Action Programs," Urban Affairs Quarterly, 3 (1967), 34-45.

Singer, I. A. "An Objective Method for Site Evaluation," Proceedings of the fifty-second annual meeting of the Air Pollution Control Association, 1959.

Haagen-Smit, A. J. "The Control of Air Pollution," Scientific American 210 (1964), 24-31.

Stanford Research Institute. "Future Urban Transportation Systems-- Description, Evaluation, and Programs." Final Report, Clark Henderson, 1968.

Stanley, W., and A. N. Heller. "Air Resources Management in the Chicago Metropolitan Area," Journal of the Air Pollution Control Association 16 (1966), 536-40.

Taffe, C. "Air Pollution Control Programs Grants--The First Year of Experience," Journal of the Air Pollution Control Association, 15 (1965), 403-8.

Ulbrich, E. A. "Adapredictive Air Pollution Control for the Los Angeles Basin," Socio-Economic Planning Sciences, 1:3 (July, 1968), 423-4

Wayne, Lowell G. "Implementation of an Air Pollution Program." Allan Hancock Foundation, University of Southern California, 1964.

Williams, J. S. Air Pollutant Emissions Related to Land Area--A Basis for a Preventative Air Pollution Control Program. Arlington, Va.: National Air Pollution Control Administration, Department of Health, Education and Welfare.

Wood, S., and D. Lembke. The Federal Threats to the California Landscape. Sacramento: California Tomorrow, 1967.

Wronske, W., E. W. Anderson, A. D. Berry, A. P. Bernhart, and H. A. Belyea. "Air Pollution Considerations in the Planning and Zoning of a Large Rapidly Growing Municipality," Journal of the Air Pollution Control Association 16:3 (March, 1966), 157-58.

Zimmerman. J. F. "Political Boundaries and Air Pollution Control," Journal of Urban Law, 46 (1968), 173.

POLITICAL SCIENCE

Abrams, Charles. "Foreward." Redoing America. Edmund Faltermayer. New York: Harper and Row, 1968.

Achenbach, P. R. "The City: A Challenge to Engineering and Political Sciences," ASHRAE, 11:3 (March, 1969), 33-38.

"Air Pollution," Federal Aid Reporter (January-February, 1967), 1-9.

Air Pollution Control. Community Action Guide for Public Officials: No. 5, Staffing and Financing. National Association of Counties Research Foundation, 1967. Washington, D.C.

"Air Pollution: Who Will Make the Decisions?" Safety Maintenance, 138:1 (July, 1969), 19-21, 51.

Anderson, James. Politics and Economic Policy Making. Reading, Massachusetts: Addison-Wesley, 1970.

Anderson, Walt. Politics and Environment: A Reader in Ecological Crisis. Pacific Palisades, California: Goodyear, 1970.

Auerbach, Irwin, and K. Flieger. "The Importance of Public Education in Air Pollution Control," Journal of the Air Pollution Control Association, 17 (February, 1967), 102-4.

Ayres, Stephen. "The Citizens' Role in Air Pollution." Speech before the American Medical Association's National Congress on Environmental Health Management, April 26, 1967. Reprint; Washington, D.C.: U. S. Public Health Service, 1967.

Bailey, Stephen. Congress Makes a Law. New York: Columbia University Press, 1950.

Barber, James. Lawmakers. New Haven: Yale University Press, 1965.

_____. Power in Committees. Chicago: Rand McNally, 1966.

Barnes, D. P. "A Practical Relationship Between Government and Industry," Journal of the Air Pollution Control Association, 15 (October, 1965), 465-66.

Beck, James, and Marrianna Stapel. "The Role of the Public in the Control of Air Pollution." Air Pollution Project: An Educational Experiment in Self-Directed Research, Summer, 1968. Pasadena, California: Associated Students of the California Institute of Technology, 1969.

Berg, J. "Governmental Responsibility for Air Pollution Control," Proceedings of the International Clean Air Congress, London, October, 1966, pp. 281-84.

Bermingham, P. E. "Federal Government and Air and Water Pollution," Business Law, 23 (January, 1968), 467.

Bernstein, Marver. The Regulation of Business by Independent Commission. Princeton: Princeton University Press, 1955.

Berry, Phillip. "After Words, Action?" Sierra Club Bulletin, 55:2 (January, 1970).

Boffey, P. "Smog: Los Angeles Running Hard, Standing Still," Science (September 6, 1968), 990-92.

Buchwald, Art. "Another Year, Another Problem to Embrace" Los Angeles Times, March 1, 1970, sect. G, p. 7, cols. 1-2.

Cahn, Robert, quoted by. "Poll Finds Alarm Over Pollution," Christian Science Monotor, March 11, 1969, p. 13.

Caldwell, Lynton. "Biopolitics: Science, Ethics, and Public Policy," Yale Review, 54 (Autumn, 1964), 1-16.

_____. "Environment: A New Focus for Public Policy," Public Administration Review, 23 (1963), 132-40.

_____. ed. Environmental Studies. Vol. 1, Bloomington: Indiana University, Institute of Public Administration, 1967.

_____. "Public Policy Implications of Environmental Control." Social Sciences and the Environment, Morris Garnsey and James Hebbs, eds. Boulder: University of Colorado Press, 1967.

Cambell, Angus, Philip Converse, Warren Miller, and Donald Stokes. American Voter. New York: Wiley, 1960.

Cartwright, Dorwin. "Public Opinion Polls and Democratic Leadership." Public Opinion and Propaganda. D. Katz, D. Cartwright, Samuel Eldersveld, and A. Lee, eds. New York: Holt, Rinehart and Winston, 1964.

Chambers, Leslie. "Classification and Extent of Air Pollution Problems." Air Pollution. Second edition. Volume 1. Arthur Stern., ed. New York: Academic Press, 1968.

Clapp, Charles. The Congressman: His Work as He Sees It. New York: Doubleday Anchor, 1963.

Congressional Quarterly Weekly Report, 25 (May 5, 1967), 723-26.

Conine, Ernest. "Improve Environment or Lives," Los Angeles Times, March 29, 1970, Sect. G, p. 7, cols. 3-5.

Cox, Ernie. "A Plea for Tougher Smog Laws," Oakland Tribune, July 13, 1969, p. 25, col. 2.

Crowe, Jay. "Toward a Definitional Model fo Public Perceptions Toward Air Pollution, Journal of the Air Pollution Control Association, 18 (March, 1968), 154-57.

Dahl, Robert. Preface to Democratic Theory. Chicago: University of Chicago Press, 1956.

Davies, J. C., III. The Politics of Pollution. New York: Pegasus, 1970.

DeFleur, Melvin, and Otto Larsen. The Flow of Information. New York: Harper, 1958.

Degler, S. E., and S. C. Bloom. "Federal Pollution Control Programs: Water, Air, and Solid Wastes." Environmental Management Series. Washington, D.C.: Bureau of National Affairs, 1969.

De Groot, I. "People and Air Pollution: A Study of Attitudes in Buffalo, New York," Journal of the Air Pollution Control Association, 16 (May, 1966), 245-47.

_____. Also see Sociology Section.

Dixon, J. P. "Public Policy in Air Conservation," Archives of Environmental Health, 8 (January, 1964), 1-18.

Dreisbach, Robert. Handbook of the San Francisco Region. Stanford: Stanford University Press, 1970.

Dubos, Rene. "We Can't Buy Our Way Out," Psychology Today, 3 (March, 1970), 20-22, 86-87.

Easton, David. "The New-Revolution in Political Science," American Political Science Review (December, 1969), 1051-61.

_____. The Political System. New York: Knopt, 1959.

Eckstein, Harry. Pressure Group Politics. Stanford: Stanford University Press, 1964.

Edelman, Murray. Symbolic Uses of Politics. Urbana, Illinois: University of Illinois Press, 1964.

Edelman, S. "Federal Air and Water Control: Application of the Commerce Power to Abate Interstate and Intrastate Pollution," George Washington Law Review, 33 (June, 1965), 1067-87.

_____. "The Law of Federal Air Pollution Control," Journal of the Air Pollution Control Association, 16 (October, 1966), 523-25.

Environmental Quality Study Council: Progress Report. Sacramento: State of California, February, 1970

Esposito, John. "Air and Water Pollution: What to Do While Waiting for Washington," Harvard Civil Rights--Civil Liberties Law Review, 5:1 (January, 1970), 32-52.

_____. Vanishing Air. New York: Grossman, 1970.

Ewald, William. Environment and Change and Environment and Policy. Bloomington, Indiana: Indiana University Press, 1968.

Faltermayer, E. Redoing America. New York: Harper and Row, 1968.

Fenno, Richard. The Power of the Purse. Boston: Little, Brown, 1967.

First Annual Report of the Council on Environmental Quality. Washington, D.C.: August, 1970.

Froman, L. The Congressional Process: Strategies, Rules, and Procedures. Boston: Little, Brown, 1967.

Grantham, J. N. "Keeping the (U.S.) Air Clean," Foreign Trade, 131 (January, 1969), 27-28.

Goulding, C. "Definition of Air Quality Control Regions: Approach and Experience to Date," Journal of the Air Pollution Control Association, 18 (September, 1968), 591-95.

Hagevik, G. "Decision Processes in Air Pollution Control." Paper presented at the sixty-first annual meeting, Air Pollution Control Association, St. Paul, June, 1968. Paper No. 68-114.

Harris, Louis. "New Priorities on Spending." Washington, D.C.: Washington Post Co., 1967.

Jacob, Charles. Policy and Bureaucracy. Princeton: Van Nostrand, 1966.

Jennings, M. Kent. Community Influentials. New York: Free Press, 1964.

Jewell, Malcolm, and Samual Patterson. The Legislative Process in the United States. New York: Random House, 1966.

Keefe, W., and M. Ogul. The American Legislative Process: Congress and the States. Englewood Cliffs: Prentice-Hall, 1968.

Key, V. O. Politics, Parties, and Pressure Groups. Fifth edition. New York: Crowell, 1964.

_____. Public Opinion and American Democracy. New York: Knopf, 1961.

Kohlmeier, Lewis. The Regulators. New York: Harper and Row, 1969.

Kovtiz, Ray. "Gaining Public Acceptance for California's Auto Smog Control Program," Journal of the Air Pollution Control Association, 17 (January, 1967), 26-27.

Krislov, S., and L. Muslof. The Politics of Regulation. Boston: Houghton Mifflin, 1964.

Lane, R. Political Life. New York: Free Press, 1959.

Lang, D., and G. E. Lang. Politics and Television. Chicago: Quadrangle Books, 1968.

Leduc, E. C. "The Socio-Political Characteristics of Urban Governments Engaged in Air Pollution Control Activities," Journal of the Air Pollution Control Association, 18:11 (November, 1968), 733-37.

Levin, E. "Smog, Cars, and the Auto Oil Complex," <u>ADA World Magazine</u> (February, 1970).

Lewis, Howard. <u>With Every Breath You Take</u>. New York: Crown, 1965.

Lieber, H. "Controlling Metropolitan Pollution Through Regional Airsheds: Administrative Requirements and Political Problems," <u>Journal of the Air Pollution Control Association</u>, 18 (February, 1968), 86-93.

Lindblom, Charles. <u>The Intelligence of Democracy</u>. New York: Free Press, 1965.

_____. <u>The Policy Making Process</u>. Englewood Cliffs: Prentice-Hall, 1968.

Linton, R. M. "What We Must Do--Politically and Socially--To Restore the Environment," <u>American Journal of Public Health</u>, 68 (November, 1968), 2055.

Los Angeles County Air Pollution Control District. <u>Profile of Air Pollution Control in Los Angeles County</u>. 1969.

Loveridge, Ronald O. "Air Pollution and the Public Will," <u>California Air Environment</u> (April-June, 1969), 5.

_____. "Socio-Political Constraints on Establishing Policies for a Habitable Environment." Paper presented at the First National Symposium on Habitability, Los Angeles, May 11-14, 1970, pp. 2-

_____. "Types, Ranges, and Methods for Classifying Human Behavioral Responses to Air Pollution." <u>Development of Air Quality Standards</u>. Arthur Atkisson and Richard Gaines, eds. Columbus, Ohio: Merrill, 1970.

Lowi, T. <u>The End of Liberalism</u>. New York: Norton, 1969.

MacKenzie, V. G. <u>The Federal Role in Air Pollution</u>. Washington, D.C.: U.S. Department of Health, Education and Welfare, U.S. Government Printing Office, 1963.

Maga, John. <u>Air Resource Management in the San Francisco Bay Area</u>. Berkeley: Institute of Governmental Studies, University of California at Berkeley, 1965.

_____. "California Experience with Air Quality Standards," <u>Journal of the Air Pollution Control Association</u>, 17 (February, 1967), 107-11.

"Man's Control of the Environment," Congressional Quarterly (August, 1970).

Margolis, Jan. "Our Country 'Tis of Thee, Land of Ecology," Esquire, 73 (March, 1970), 124, 172-79.

Matthews, Donald. U. S. Senators and Their World. Chapel Hill: University of North Carolina Press, 1960.

McCloskey, Herbert. Political Inquiry. New York: Harper, 1969.

Meggonnell, William. "Developing Abatement Policies Under the Clean Air Act," Journal of the Air Pollution Control Association, 16 (May, 1966), 254-56.

Meredith, H. H. "Platitudes or Performance," Journal of the Air Pollution Control Association, 16 (October, 1966), 547-49.

Milbrath, Lester. Political Participation. Chicago: Rand McNally, 1965.

"Mr. Hickel Has a Choice," Sierra Club Bulletin (July, 1969), p. 2.

Mitchell, Joyce, and William Mitchell. Political Analysis and Public Policy. Chicago: Rand McNally, 1969.

Moynihan, Daniel. "Crises of Confidence," The Public Interest, (Spring, 1967), pp. 3-10.

Muskie, Edmund S. "Environmental Jurisdiction in the Congress and the Executive," Maine Law Review, 22 (1970), 171.

N.A.P.C.A., Air Quality Criteria for Particulate Matter, National Air Pollution Control Administration, U.S. Public Health Service, Washington, D.C. (January, 1969), p. 7-13.

Nimmo, D., and T. Ungs. American Political Patterns. Boston: Little, Brown, 1967.

Pastier, John. "Conservation Group Wins Significant Legislative Battles," Los Angeles Times, September 28, 1969, sect. J, p. 15, col. 1.

Perman, Robert. "Emerging Concepts of Air Pollution Problems as Seen by the Political Scientist," Journal of the Air Pollution Control Association, 16:10 (October, 1966), 532-35.

"Pollution Battle Plan," Newsweek (February 23, 1970, p. 24, col. 1.

Pranger, Robert. The Eclipse of Citizenship. New York: Holt, Rinehart, 1968.

Project 70, A Clean Air Environment for California. A plan of action presented to the California State Legislature by the University of California. January 2, 1970.

Ranney, Austin. Political Science and Public Policy. Chicago: Markham, 1968.

Rapoport, Roger. cited in "Los angeles Has a Cough," Esquire (July, 1970), 84.

Reagan, Michael. "Administrative and Legal Dimensions of Air Pollution Control." Proposed Applied Research and Development Projects. Supplement No. 1 to Project 70, A Clean Air Environment for California. Riverside, California: January 23, 1970, pp. 3-4.

_____. Science and the Federal Patron. New York: Oxford, 1969.

Redford, Emmette. Administration of National Economic Control. New York: MacMillan, 1952.

_____. Democracy in the American State. New York: Oxford, 1969.

Ridker, Ronald. "Strategies for Measuring the Cost of Air Pollution." The Economics of Air Pollution. Harold Wolozin, ed. New York: Norton, 1966.

Ripley, Randall. Public Policies and Their Politics. New York: Norton 1966.

Rosenau, James. Public Opinion and Foreign Policy. New York: Random House, 1961.

Rourke, Francis. Bureaucracy, Politics, and Public Policy. Boston: Little, Brown, 1969.

Rumford, W. B. "The Politics of Pollution," Journal of the Air Pollution Control Association, 16 (July, 1966), 359-61.

Sacco, J., and E. Leduc. "An Analysis of the Effect of Social, Economic and Political Factors on State Air Pollution Expenditures." Paper presented at the sixty-first annual conference, Air Pollution Control Association, June, 1968.

Saloma, John. Congress and the New Politics. Boston: Little, Brown, 1969.

Samuels, Sheldon. "The Role of Behavioral Research in Air Pollution Control." Paper delivered at the Symposium on the Role of Perceptions and Attitudes in Decision-Making in Resources Management," University of Victoria, British Columbia, April 13, 1970, pp. 23-24.

Schueneman, Jean. See Sociology Section.

Schusky, Jane. "Public Awareness and Concern with Air Pollution in the St. Louis Metropolitan Area," Journal of the Air Pollution Control Association, 16 (February, 1966), 72-76.

_____, et al. Also see Sociology Section, entry under Loring, W. C.

Seldes, Gilbert. "How Can We Get Action for Cleaner Air Through Public Communications?" Proceedings of the National Conference on Air Pollution. Washington, D.C.: Division of Air Pollution, Public Health Service, 1963. Pub. No. 10022.

Seymour, W. N. "Cleaning Up Our City Air: Proposals for Combating Air Pollution Through Affirmative Government Action Programs," Urban Affairs Quarterly, 3 (September, 1967), 34-45.

Sharkansky, I. Policy Analysis in Political Science. Markham, 1970.

Stalker, W. W., and C. Robinson. "A Method for Using Air Pollution Measurements and Public Opinion to Establish Ambient Air Quality Standards," Journal of the Air Pollution Control Association, 17 (March, 1967), 142-144.

Stein, Jane. "Priorities in Pollution: The SST and the Smogless Car," The Washington Monthly (February, 1969).

Sundquist, James. Politics and Policy. Washington, D.C.: Brookings Institution, 1968, pp. 506-37.

Train, Russell, cited by Rudy Abramson. Los Angeles Times, February, 15, 1970, sect. G, p. 1, col. 3.

Truman, David. Governmental Process. New York: Knopf, 1951.

Verba, S. Small Groups and Political Behavior. Princeton: Princeton University Press, 1961.

_____, et al. "Public Opinion and the War in Vietnam," American Political Science Review (June, 1967), 317-33.

Wahlke, John., H. Eulau, W. Buchanan, and L. Ferguson. The Legislative System. New York: Wiley, 1962.

Weissman, R. "Environmental Education: An Interview with Senator Gaylord Nelson," Arizona Teacher, 58:4 (March, 1970), 6-8, 45.

Wildavsky, Aaron. "Aesthetic Power or the Triumph of the Sensitive Minority over the Vulgar Mass: A Political Analysis of the New Economic," Daedalus (Fall, 1967), 1115-29.

_____. The Politics of the Budgetary Process. Boston: Little, Brown, 1964.

Wislon, James, Q. Varieties of Police Behavior. Cambridge: Harvard University Press, 1968.

_____. "We Need to Shift Focus." Urban Government: A Reader in Administration and Politics. Edward Banfield, ed. Rev. ed. New York: Free Press, 1969.

Woll, Peter. American Bureaucracy. New York: Norton, 1963.

Wolman, A. "Air Pollution: Time for Appraisal," Science (March 29, 1968), 1437-40.

Zeigler, H. Interest Groups in American Society. Englewood Cliffs: Prentice-Hall, 1964.

PSYCHOLOGY

Alpatov, I. M. "A Study of Gasious Ammonia Toxicity," Gigiena Truda Professora Zabolevaniia, Moscow, 8 (1964), 14-18.

Andreescheva, N. G. "Substantiation of the Maximum Permissable Concentration of Nitrobenzene in Atmospheric Air," Hygiene and Sanitation (English translation of Gigiena i Sanitariia, Moscow), 29 (1964), 4-9.

_____. "The Effects of Certain Aromatic Hydrocarbons in the Air," Hygiene Sanitation (English translation of Gigiena i Sanitariia, Moscow), 33 (1968), 13-17.

Anonymous. "Lead in the Environment and Its Effects on Humans." Berkeley: Bureau of Air Sanitation and Air and Industrial

Hygiene Labs, California State Department of Public Health, 1967.

Barnes, J. M. "Mode of Action of Some Toxic Substances (with Special Reference to the Effects of Prolonged Exposure)," British Medical Journal, 5260 (1961), 1097-1104.

Bazmadzhiero, K., G. Kurchatova, E. Davidkova, and I. Tsuetanøv. "Combined Effect of Sulfur Dioxide and Carbon Monoxide in the Atmosphere," Hygiene and Sanitation (English translation of Gigiena i Sanitariia, Moscow), 33 (1968), 81-86.

Beard, R. R. "Toxicological Appraisal of Carbon Monoxide," Journal of the Air Pollution Control Association, 19 (1968), 722-32.

_____, and G. A. Wertheim. "Behavioral Impairment Associated with Small Doses of Carbon Monoxide," American Journal of Public Health, 57 (November, 1967), 2012-22.

Beckman, A. L., and J. S. Eisenman. "Responsiveness of Temperature-Sensitive Neurons to Microelectrophoretically Applied Amines in Rates and Cats," Federation Proceedings, 29 Abs., (1970), 523.

Beliles, R. P., R. S. Clark, P. R. Belluscio, C. L. Yuile, and L. J. Leach. "Behavioral Effects in Pigeons Exposed to Mercury Vapor at a Concentration of 0.1 mg/m^3," American Industrial Hygiene Association Journal, 28 (1967), 482-84.

Berlin, M., J. Fazackerley, and G. Nordberg. "The Uptake of Mercury in the Brains of Mammals Exposed to Mercury Vapor and to Mercuric Salts," Archives of Environmental Health, 18 (1969), 719-29.

Bevilacqua, D. M., and C. W. LaBelle. "Synergistic Effects of Aerosols," American Industrial Hygiene Association Journal, 24 (1963), 448-52.

Biersteker, K., and H. de Graff. "The Determination of Maximum Acceptable Concentrations of Air Pollution," Tijdschrift voor Sociale Geneeskunde (Assen), 43 (1965), 526-28.

Bocca, F., and M. N. Battiston. "Odour Perception and Environment Conditions," Acta Oto-Laryngol (Stockholm), 57 (1964), 391-400.

Braginskaya, L. L., and V. A. Polyanskii. "The Combined Effect of Toxic Substances and Physical Stress on the Performance Capacity and Energy Metabolism in the Muscle and Liver of

Albino Mice," Gigiena Truda Professora Zabolevaniia, Moscow, 12 (1968), 46-50.

Buchberg, H., M. H. Jones, K. G. Lindh, and K. W. Wilson. "Correlation Studies of Interacting Atmospheric Variables and Eye Irritation Threshold Pt. in Air Pollution Studies with Simulated Atmospheres." Los Angeles: Department of Engineering, University of California at Los Angeles. Report No. 61-44. Vol. 1, pp. 1-107.

_____. "Studies of Interacting Atmospheric Variables and Eye Irritation Thresholds," International Journal of Air and Water Pollution (London), 7 (1963), 257-80.

Bushtueva, K. A. "The Effect of Absorption of Oxides of Sulphur," Hygiene and Sanitation (English translation of Gigiena i Sanitariia, Moscow), 29 (1964), 7-12.

Campbell, K. I., L. O. Emik, G. L. Clarke, and R. L. Plata. "Inhalation Toxicity of Peroxyacetyl Nitrate Depression of Voluntary Activity in Mice," Archives of Environmental Health, 20 (1970), 22-27.

Carson, D. "Environmental Stress and the Urban Dweller," Michigan Mental Health Research Bulletin, 2:4 (1968), 5-12.

Detrie, J. P. "La Pollution Atmospherique" Dunod, (Paris), 1969.

Dmitrieva, N. V. "Maximum Permissible Concentration of Tetrachloroethylene in Factory Air," Hygiene and Sanitation (English translation of Gigiena i Sanitariia, Moscow), 31 (1966), 387-92.

Dubrovskaya, F. I., and I. P. Lukina. "Effect of Low Concentrations of Valeric Acid Vapor on Experimental Animals," Hygiene and Sanitation (English translation of Gigiena i Sanitariia, Moscow), 31 (1966), 193-96.

Eliseeva, O. V. "Substantiation of the Maximum Permissible Carbon Dioxide Concentration in the Air of Houses and Public Buildings," Hygiene and Sanitation (English translation of Gigiena i Sanitariia, Moscow), 29 (1964), 10-15.

Emik, L. O., and R. L. Plata. "Depression of Running Activity in Mice by Exposure to Polluted Air," Archives of Environmental Health, 18 (1969), 574-79.

Epstein, D. "Detection and Prevention of Air Pollution in the U.S.S.R.," Pollution Atmosphérique (Paris) 8 (1966), 273-83.

Feldberg, W. "The Monamines of the Hypothalamus as Mediators of Temperature Responses." Recent Advances in Pharmacology. Fourth Edition. J. M. Robson and R. S. Stacey, eds. Boston: Little, Brown, 1968, 349-97.

Gofmekler, V A. "Effect of Acetaldehyde on Reflexes," Hygiene Sanitation (English translation of Gigiena i Sanitariia, Moscow), 33 (1968), 109-10.

Goldberg, H. G. and M. N. Chappell. "Behavioral Measure of Effect of Carbon Monoxide on Rats," Archives of Environmental Health, 14 (1967), 671-77.

Goldberg, M. E., H. E. Johnson, U. C. Pozzani, and H. F. Smythe, Jr. "Behavioral Response of Rats During Inhalation of Trichloroethylene and Carbon Disulfide Vapours," Acta Pharmacologica et Toxicologica 21 (1964), 36-44.

Goldsmith, J. R. "Bases and Criteria for Air Quality Standards," Journal of the Air Pollution Control Association, 14 (1964), 22-26.

──────. "Nondisease Effects of Air Pollution," Environmental Research 2 (1969), 93-101.

Goluber, A. A. "Effect of Certain Industrial Irritants on Change of Pupil Diameter in the Rabbit," Gigiena Truda Professora Zabolevaniia, Moscow, (1969), 58-59.

Gusev, M. I., A. I. Svechnikova, I. S. Dronov, M. D. Grebenskova, and A. I. Golovnia. "Determination of the Daily Average Maximum Permissible Concentration of Acrolein in the Atmosphere," Hygiene and Sanitation (English translation of Gigiena i Sanitariia, Moscow), 31 (1966), 8-13.

Hamming, M. J., and R. D. MacPhee. "Relationship of Nitrogen Oxides in Auto Exhaust to Eye Irritation--Further Results of Chamber Studies," Atmospheric Environment, 1 (1957), 577-84.

Henschler, D., S. Stier, H. Beck, and W. Neumann. "Olfactory Threshold of Some Important Irritant Gases (Sulfur Dioxide, Ozone, Nitrogen Dioxide) and Their Effect on Man When Exposed to Low Concentrations," Archiv für Gewerbepathologie und Gewerbehygiene 17 (1960), 547-70.

Hore, T., and D. E. Gibson. "Ozone, Exposure and Intelligence Tests," Archives of Environmental Health, 17 (1968), 77-79.

Horvath, M. "Effects of Toxic Compounds on Certain Central Functions," Journal of Occupational Medicine, 1 (1959) 463.

Hueter, F. G., G. L. Conther, K. A. Busch, and R. G. Hinners. "Biological Effects of Atmospheres Contaminated by Auto Exhaust," Archives of Environmental Health, 12 (1966), 553-60.

Hughes, J. R., and J. A. Mazurowski. "Studies on the Supracallosal Mesial Cortex of Unanesthetized, Conscious Mammals: Part II, The Monkey; Responses from the Olfactory Bulb," Electroencephalography and Clinical Neurophysicology (Amsterdam), 14 (1962), 635-45.

Ito, K. "Effects of Exhaust Gas on the Human Body," Japanese Air Cleaning Association, Kuki Seijo (Tokyo), 5 (1968), 27-32.

Jacobziner, H. "Lead Poisoning in Childhood: Epidemiology, Manifestations, and Prevention," Clinical Pediatrics, 5 (1966), 277-86.

Jaffe, L. S. "The Biological Effects of Ozone on Man and Animals," American Industrial Hygiene Society Journal, 28 (1967), 267-77.

_____. "The Biological Effects of Photochemical Air Pollution on Man and Animals," American Journal of Public Health, 57 (1967), 1269-77.

Kailin, E. W. "Acute Cerebral and Pulmonary Effects of Air Pollution in Washington, D.C.," Medical Annals of the District of Columbia, 37 (1968), 270-71.

Kato, K. "Ions in the Air," Japanese Air Cleaning Association, Tokyo, 2 (1964), 48-50.

Komura, S. "Electroencephalographic Studies of Carbon Monoxide Poisoning in Rabbits," Japanese Journal of Legal Medicine, Tokyo, 21 (1967), 25-48.

Korneev, Y. E. "Effect of the Combined Presence of Low Concentrations of Phenol and Acetophenone in the Urban Atmosphere," Gigiena i Sanitariia, Moscow, 30 (1965), 336-45.

Kramarenko, I. B., and F. I. Grishko. "A Study into Long-Term Action of Carbon Disulfide upon the Function of the Neuromuscular System in Animals of Different Age," Gigiena Truda Professora Zabolevaniia, Moscow, 13 (1969), 33-35.

Kulakov, A. E. "The Effect of Small Concentrations of Hexamethylenediamine on Man," Gigiena Truda Professora Zabolevaniia, Moscow, 29 (1964), 8-13.

Lagerwerff, J. M. "Prolonged Ozone Inhalation and Its Effects on Visual Parameters," Aerospace Medicine, 34 (1963), 479-86.

Levine, B. S. "U.S.S.R. Literature on Air Pollution and Related Occupational Diseases." Washington, D.C.: U. S. Public Health Service, Department of Health, Education, and Welfare, 1964.

Lewis, T. R., F. G. Hueter, and K. A. Busch. "Irradiated Automobile Exhaust: Its Effects on the Reproduction of Mice," Archives of Environmental Health, 15 (1967), 26-35.

MacDonald, W. E., W. B. Deichmann, and E. Bernal. "Subacute Effects of Vapors of Nitro-Olefins upon Experimental Animals," American Industrial Hygiene Association Journal, 24 (1963), 539-40.

Makhinya, A. P. "Hygienic Assessment of Atmospheric Air Polluted by Sulfur Dioxide with Phenol," Hygiene and Sanitation (English translation of Gigiena i Sanitariia, Moscow), 31 (1969), 314-17.

Medalia, N. Z. "Air Pollution as a Socio-Environmental Health Problem--A Survey Report," Journal of Health and Human Behavior, 5 (1964), 154-65.

Minaev, A. A. "Determination of the Maximum Permissible Concentration of a-methylstyrene Vapor in the Atmosphere," Hygiene and Sanitation (English translation of Gigiena i Sanitariia, Moscow), 31 (1969), 157-61.

Murphy, S. D. "A Review of Effects on Animals of Exposure to Auto Exhaust and Some of its Components," Journal of the Air Pollution Control Association, 14 (1964), 303-8.

Orcutt, J. A. "The Quantal Response in Environmental Toxicology. Part I--The Measurement of Eye Irritation as a Quantal Response for Correlation with Aerometric Data from Polluted Atmospheres," Journal of the American Osteopathic Association, 66 (1967), 1376-83.

Owens, E. J., and C. L. Punte. "Human Respiratory and Ocular Irritation Studies Utilizing o-chlorobenzylidene Malononitrile Aerosols: Effect of Particle Size," American Industrial Hygiene Association Journal, 24 (1963), 262-64.

Paris, J. "Study of the Excitation-Duration Curves After Acute Occupational Intoxication from Carbon Monoxide," Rassegna Medicine Industriale, Rome 33 (1964), 275-91.

Partsef, D. P. "Chronic Effect of Certain Components of Exhaust Gases from Motor Cars," Hygiene and Sanitation (English translation of Gigiena i Sanitariia, Moscow), 31 (1966), 363-68.

Pazynich, V. M. "Maximum Permissible Concentration of Vanadium Pentoxide in the Atmosphere," Hygiene and Sanitation (English translation of Gigiena i Sanitariia, Moscow), 31 (1966), 6-11.

Perlstein, M. A., and R. Attala. "Neurologic Sequelae of Plumbism in Children," Clinical Pediatrics, 5 (1966), 292-98.

Pogosian, U. G. "The Effect on Man of the Combined Action of Small Concentrations of Acetone and Phenol in the Atmosphere," Hygiene and Sanitation (English translation of Gigiena i Sanitariia, Moscow), 30 (1967), 1-9.

_____. "The Joint Action of Small Concentrations of Acetone and Phenol in the Atmosphere on Man," Gigiena i Sanitariia, Moscow, 30 (1965), 3-9.

Prince-Epstein, D. "Studies on Atmospheric Pollution in the Soviet Union," Bull. Inst. Nat. Sante Rech. Med., Paris, 23 (1968), 63-82.

Rall, D. P. "Difficulties in Extrapolating the Results of Toxicity Studies in Laboratory Animals to Man," Environmental Research, 2 (1969), 360-67.

Reynolds, R. W. "The Use of Reaction Time in Monkeys for the Study of Information Processing." Animal Psychophysics. W. C. Stebbins, ed. New York: Appleton-Century-Crofts, 1970.

_____, and R. R. J. Chaffee. "Studies on the Combined Efforts of Ozone and a Hot Environment on Reaction Time in Subhuman Primates." University of California, Project Clean Air. Research reports, Volume 2, 1970.

Rjazanov, V. A. "Criteria and Methods for Establishing Maximum Permissible Concentrations of Air Pollution," Bulletin of the World Health Organization, 32 (1965), 389-98.

Ross, P. H. Why Families Move: A Study in the Social Psychology of Urban Residential Mobility. Glencoe, Free Press, 1965.

Ruffin, J. B. "Functional Testing for Behavioral Toxicity: A Missing Dimension in Experimental Environmental Toxicology," Journal of Occupational Medicine, 5 (1963), 117-21.

Sadilova, M. S., K. P. Seliankina, and O. K. Shturkina. "The Effect of Hydrogen Flouride on the Central Nervous System in an Experiment," Gigiena Truda Professora Zabolevaniia, Moscow, 30 (1965), 11-15.

Saito, K., and S. Abe. "Relation Between Lead Poisoning and Electroencephalography," Japan Journal of Industrial Health, Tokyo, (1965), 20-27.

Sakhovs'ka, N. M. "Disturbances in Conditioned Reflexes of Rats During Experimental Intoxication by Hydrogen Sulfide-sulfurous Anhydrite Complex," Fiziologichnii Zhurnal Academica Nauk Ukrainskoi RSR, Kiev, 13 (1967), 521-26.

Schlipkoter, H. W. "Effects of Air Pollution on Man." Air Pollution as a Sociological Problem. A. Silbermann, Lufthyg., In Dortmund Vortragstag, 20 (1964), 26-34.

Schulte, J. H. "Effects of Mild Carbon Monoxide Intoxication," Archives of Environmental Health, 7 (1963), 524-37.

Schusky, J., L. Golner, S. Z. Mann, and W. C. Loring. "Methodology for the Study of Public Attitudes Concerning Air Pollution," Journal of the Air Pollution Control Association, 14 (November, 1964), 445-48.

Shalamberidze, O. P. "Reflex Effects of Mixtures of Sulfur and Nitrogen Dioxides," Hygiene and Sanitation (English translation of Gigiena i Sanitariia, Moscow), 32 (1967), 10-15.

Shulga, T. M. "Data to Substantiate the Maximum Permissible Daily Average Concentration of Carbon Monoxide in the Atmospheric Air," Hygiene and Sanitation (English translation of Gigiena i Sanitariia, Moscow), 30 (1965), 3-6.

Smith, S. M. "How Is Air Pollution Affecting Esthetic Values in the United States?" Proceedings of the National Conference on Air Pollution, 3 (1966), 545-53.

Solomin, G. I. "Experimental Data on the Hygenic Substantiation of the Single Permissible Concentrations of Isopropylbenzene and Isopropylbenzene Hydroperoxide in Atmospheric Air," Hygiene and Sanitation (English translation of Gigiena i Sanitariia, Moscow), 29 (1964), 1-8.

Stern, A. C. "Summary of Existing Air Pollution Standards," Journal of the Air Pollution Control Assocation, 14 (1964), 5-15.

Stoefen, D. "The Foundation of the Soviet MAC Value for Lead," Archiv für Hygiene und Bakteriologie Munich, 152 (1968), 93-96.

Stokinger, H. E. "Suggested Principles and Procedures for Developing Data for Threshold Limit Values for Air: Experimental Animal Data and Data From Human Subjects." Industrial Hygiene Foundation of America, Inc., Chemical-Toxicological Series, Bulletin No. 8-69, Pittsburgh, 1969.

Stone, H. "Influence of Temperature on Olfactory Sensitivity," Journal of Applied Physiology, 18 (1963), 746-51.

Stopps, G. J., and M. McLaughlin. "Psychophysiological Testing of Human Subjects Exposed to Solvent Vapors," American Industrial Hygiene Society Journal, 28 (1967), 43-50.

Teichner, W. H. "An Exploration of Some Behavioral Techniques for Toxicity Testing," Journal of Psychology, 65:1 (1967), 69-90.

Tepikina, L. A. "Biological Effects of Mesidine as an Atmospheric Pollutant," Hygiene and Sanitation (English translation of Gigiena i Sanitariia, Moscow), 33 (1968), 299-302.

Tkach, N. Z. "Combined Effect of Aceton and Acetophenone in the Atmosphere," Gigiena Truda Professora Zabolevaniia, Moscow, (transl. by JPRS), 30 (1965), 179-85.

Tkachenko, Z. A., A. N. Tishchenko, A. T. Zatsepilin, and E. Z. Dimitrova. "External Breathing in Workers under Conditions of Slightly Elevated Carbon Monoxide Concentration," Verachebnoe Delo (Kiev), 12 (1966), 78-81.

Trakhtenberg, I. M. "The Problem of Micromercurialism in the Light of New Experimental Facts and Observations," Gigiena Truda Professora Zabolevaniia, Moscow, 27 (1962), 17-25

Truhaut, R. "Tolerable Limits for Toxic Materials in Industry: Divergences and Points of Agreement at the International Level," Archives des Maladies Professionelles de Médicine du Travail et de Sécurite Sociale, Paris, 26 (1965), 41-56.

Ubaidullaev, R. "Combined Effect of Low Concentrations of Furfurol, Methanol and Hydrolysis Ethanol Under Experimental Conditions,"

Hygiene and Sanitation (English translation of Gigiena i Sanitariia, Moscow), 32 (1967), 313-19.

Vernon, R. J., and R. K. Ferguson. "Effects of Trichloroethylene on Visual-Motor Performance," Archives of Environmental Health, 18 (1969), 894-900.

Von Post-Lingen, M. L. "The Significance of Exposure to Small Concentrations of Carbon Monoxide (Results of an Experimental Study on Healthy Persons)," Proceedings of the Royal Society of Medicine 57 (1964), 1021-29.

Wayne, L. G. "Eye Irritation as a Biological Indicator of Photochemical Reactions in the Atmosphere," Atmospheric Environment, London, 1 (1967), 97-104.

Weitzman, D. O., J. A. S. Kinney, and S. M. Luriator. "Effect on Vision of Repeated Exposure to Carbon Dioxide." Groton, Conn.: Naval Submarine Medical Center, Submarine Medical Research Laboratory, 1969.

William, T. F. "Air Pollution and the Public Conscience," Journal of Health and Human Behavior, 1966.

Wohlwill, J. "The Emerging Discipline of Environmental Psychology," American Psychologist, 25:4, 303-11.

Wolstenholme, G., and R. Porter, eds. Drug Responses in Man. Boston: Little, Brown, 1967.

Xintaras, C. "Behavioral Toxicology Looks at Air Pollutants," Environmental Science and Technology, 2 (1968), 731-73.

Yaakmees, V. A. "The Establishment of the Maximum Permissible Concentration of Shale Gasoline in the Atmosphere," Hygiene and Sanitation (English translation of Gigiena i Sanitariia, Moscow), 31 (1966), 295-301.

SOCIOLOGY

"Air Pollution as a Private Nuisance," Washington and Lee Law Review, 24 (Fall, 1967), 314.

"Air Pollution Control: Gaining Community Support," American County Government, 32 (January, 1967), 27-42.

Alinsky, S. Reveille for Radicals. Chicago: University of Chicago Press, 1946.

American Association for the Advancement of Science. Air Conservation. Washington, D.C., 1965. Publication No. 80, pp. 273-81.

Anonymous. (a) "Gallup: 73% of Americans Will Pay More Taxes," National Wildlife (April-May, 1969), 18-19.

_____. (b) "The View From the Pinnacle," Fortune (September, 1969), 93-94.

_____. (a) "New Conservation Poll," National Wildlife (December-January, 1970), 18-19.

_____. (b) "How Students See the Pollination Issue," Business Week (February 7, 1970), 86-88.

Atkinson, G. A. "Pollution, Protest and Participation," Ekistics, 26:156 (1968), 430-431.

Auerbach, I. L., and K. Flieger. "The Importance of Public Education in Air PollutionControl," Journal of the Air Pollution Control Association, 17:2 (February, 1967), 102-4.

Ayers, S. M. "The Citizen's Role in Air Pollution . . ." Presented before the American Medical Association National Congress on Environmental Health Management, U. S. Public Health Service, Washington, D.C., April 26.

Banfield, E. C. Political Influence. Glencoe: Free Press, 1961.

Barker, Roger. "On the Nature of the Environment," Journal of Social Issues (November, 1963), 18-29.

Barnett, H. "Environmental Policy and Management." Social Sciences and the Environment. Boulder, Colo.: University of Colorado Press, 1967, pp. 219-37.

Bigman, S. K. "Occupational Health and Social Research," Journal of Health and Human Behavior, 5:4 (Winter, 1964), 170-76.

Brodbelt, S. "Air Pollution: An Exercise in Problem Solving," Social Studies, 61 (February, 1970), 74-78.

Buckley, W. F. Sociology and Modern Systems Theory. Englewood Cliffs, N.J.: Prentice-Hall, 1967.

Burgess, E. "The Growth of the City: An Introduction to a Research Project." The City. R. Park, E. Burgess, and R. D. McKenzie, eds. Chicago: University of Chicago Press, 2925, pp. 47-62.

Cederlof, R., et al. "The Epidemiological Approach in Estimating MAC Values with Special Reference to Medico-Sociological and Statistical Methodology," Nordisk Hygienisk Tidskrift, Copenhagen, 46 (1965), 33-43.

Chambers, Leslie A. "National Problems of Environmental Health," Journal of Health and Human Behavior, 5:4 (Winter, 1964), 133-36.

Clinard, M. B. Slums and Community Development: Experiments in Self-Help. New York: Free Press, 1966.

Colley, J. R. T., and W. W. Holland. "Social and Environmental Factors in Respiratory Diseases," Archives of Environmental Health, 14:1 (January, 1967), 157-61

Crain, R. L., et al. The Politics of Community Conflict: The Flouridation Decision. Indianapolis: Bobbs-Merrill, 1969.

Crocker, T. D. "Some Economics of Air Pollution Control," Natural Resources Journal (April, 1968), 241.

Crowe, M. J. See Political Science Section.

de Groot, I. "Trends in Public Attitudes Toward Air Pollution," Journal of the Air Pollution Control Association, 17:10 (October, 1967), 679-81.

_____, W. Loring, A. Richm, Jr., S. D., Samuels, and W. Winkelstein, Jr. "People and Air Pollution: A Study of Attitudes in Buffalo, New York," Journal of the Air Pollution Control Association, 16 (May, 1966), 245-47.

Devall, William B. "Conservation: An Upper Middle-Class Social Movement: A Replication," Journal of Leisure Research (Spring, 1970), 123-126.

Doob, Leonard W. "Tropical Weather and Attitude Surveys," Public Opinion Quarterly, 32:3 (Fall, 1968), 423-30.

Duncan, O. D. "From Social System to Ecosystem," Sociological Inquiry (Spring, 1961), 140-49.

Dunham, H. W. "The Ecology of the Functional Psychoses in Chicago," American Sociological Review, 2 (August, 1937), 467-79.

Eckenrode, R. T. "Behavioral Aspects of Pollution Control," Consulting Engineer, 32:3 (March, 1969), 168-73.

Eckstein, M. E. "Technology, Organizations, and the Social Aspects of the Control of Air Pollution." Paper presented at the sixty-second annual meeting, Air Pollution Control Association, New York, June 22-26, 1969.

Edelman, M. The Symbolic Uses of Politics. Urbana: University of Illinois Press, 1964.

Environmental Pollution Panel, President's Science Advisory Committee Report. Restoring the Quality of our Environment. Washington, D.C.: U. S. Government Printing Office, 1965. Appendix XI, p. 43.

Firey, Walter. "Sentiment and Symbolism as Ecological Variables," American Sociological Review 10 (April, 1945), 140-48.

Flacks, Richard. "The Liberated Generation," Journal of Social Issues (December, 1967), 521-43.

Galbraith, J. K. The Affluent Society. Boston: Houghton Mifflin, 1958.

Gallup Opinion Index. Most Important Problem. Report No. 56 (Princeton, N.J.: February, 1970).

_____. Domestic Priorities of College Students. Report No. 55 (Princeton, N.J.: January, 1970).

Gargan, J. J. "The Politics of Water Pollution in New York State-- The Development and Adoption of the 1965 Pure Waters Program," Ph.D. dissertation, Syracuse University, 1968.

Garnsey, M., and J. R. Hibbs, eds. Social Sciences and the Environment. Boulder: University of Colorado Press, 1967.

Gold, D. "Public Concern and Beliefs About Air Pollution in California: A Statewide survey." Research Project S-11. Vol. 3. Riverside: University of California, Project Clean Air, 1970.

Goldsmith, J. R. "Nondisease Effects of Air Pollution," Environmental Research, 2:2 (February, 1969), 93-101.

Goodenough, W. H. "Human Problems in the Conduct of Environmental Health Programs: The Role of Environment in Culture and Society," Journal of Health and Human Behavior, 5:4 (Winter, 1964), 141-44.

Gordon, M. Sick Cities: Psychology and Pathology of American Urban Life. Baltimore: Penguin Books, 1965.

Grober, R. C. "The Problem of Air Pollution and Its Effects on Various Aspects of Society." Paper presented at the ASMA Meeting, Hartford, Conn., March 26, 1966.

Hawley, A. Human Ecology. New York: Ronald Press, 1950.

Heller, A. N. "Methods of Evaluating Socio-Economic Effects in Air Pollution." Proceedings of the International Conference on Atmospheric Emissions from Sulfate Pulping. E. R. Hendrickson, ed. Delano, Fla.: Painter, 1966, pp. 141-56.

"How Students See the Pollution Issue," Business Week (February 7, 1970), 86-88.

Hoyt, H. One Hundred Years of Land Values in Chicago. Chicago: University of Chicago Press, 1933.

Jonson, E. "Annoyance Reactions to External Environmental Factors in Different Sociological Groups," Acta Sociologica, 7:4 (1963), 228-63.

Kates, R. W., and J. F. Wohlwill, eds. "Man's Response to the Physical Environment," Journal of Social Issues, 22:4 (1966), 1-140.

King, K. "Crisis of Concern: Air and Water Pollution Solutions Sought that Recognize People and Their Values," Public Relations Journal, 23 (July, 1967), 12-14.

Kovitz, R. "Gaining Public Acceptance for California's Auto Smog Control Program," Journal of the Air Pollution Control Association 17:1 (January, 1967), 26-27.

Krutilla, J. V. "Some Environmental Effects of Economic Development," Daedalus, 96:4 (Fall, 1967), 1058-70.

Leduc, E. C. Journal of the Air Pollution Control Association, 18:11 (November, 1968), 733-37.

_____, et al. "Manpower Policies in Selected Air Pollution Control Agencies," Journal of the Air Pollution Control Association, 18:4 (April, 1968), 211-15.

Lewin, Kurt. "Group Decision and Social Change." Readings in Social Psychology. E. E. Maccoby, ed. New York: Holt, Rinehart and Winston, 1958, pp. 197-211.

Lewis, P. H., Jr. "Environment Awareness," Arts in Society, 4:3 (Fall-Winter, 1967), 477-89.

Lippitt, R., et al. The Dynamics of Planned Change. New York: Harcourt, Brace, and World, 1958.

Loomis, C., and J. A. Beegle. Rural Sociology: The Strategy of Change. Englewood Cliffs, N.J.: Prentice-Hall, 1957.

Loring, W. C. "Residential Environment: Nexus of Personal Interactions and Healthful Development," Journal of Health and Human Behavior, 5:4 (Winter, 1964), 166-69.

_____. L. Golner, S. Z. Mann, and J. Schusky. "Methodology for the Conduct of Surveys of Public Attitudes Concerning Air Pollution." Paper presented at the Fifty-seventh Annual Meeting, Air Pollution Control Association, Houston, Texas, June, 1964.

Loveridge, R. O. See Political Science Section.

Lynch, K. The Image of the City. Massachusetts: Cambridge Technology Press, 1960.

MacKenzie, V. G. "The Role of the Scientist and Citizen, A Case Study." Washington, D. C.: U. S. Department of Health, Education, and Welfare, Public Health Service, Division of Air Pollution, 1965.

Mankoff, Milton. "Power in Advanced Capitalist Society: A Review Essay on Recent Elitist and Marxist Criticism of Pluralist Theory," Social Problems, 17:3 (Winter, 1970), 418-30.

Marquis, Stewart. "Ecosystems, Societies, and Cities," American Behavioral Scientist, 11:6 (July-August, 1968), 11-15.

Maslach, G. "The Reorganization of Educational Resources," Daedalus 96:4 (Fall, 1967), 1200-09.

Means, R. L. "Sociology, Biology and the Analysis of Social Problems," Social Problems, 15:2 (Fall, 1967), 200-12.

Medalia, N. Z. "Air Pollution as a Socio-Environmental Health Problem: A Survey Report," Journal of Health and Human Behavior, 5:4 (1964), 154-65.

_____. "Citizen Participation and Environmental Health Action: The Case of Air Pollution Control," American Journal of Public Health, 59:8 (August, 1969), 1385-91.

_____, and A. L. Finkna. Community Perception of Air Quality: An Opinion Survey in Clarkston, Washington. Washington, D.C.: U. S. Public Health Service, June, 1965. Pub. No. 999-AP-10.

Megonnell, W. H. "Air Pollution Control: Its Impact Upon Municipalities, Industry and the Individual," Air Engineering, 9:7 (July, 1967), 12-14.

Merelman, R. M. "On the Neo-Elitist Critique of Community Power," The American Political Science Review, 62:2 (June, 1968), 451-60.

Merton, R. K. Social Theory and Social Structure. Glencoe, Illinois: Free Press 1957. Chapter 8.

Meyerson, Martin, and E. C. Banfield. Politics, Planning and the Public Interest. New York: Free Press, 1955.

Michelson, W. Man and His Urban Environment. Reading, Massachusetts: Addison-Wesley, 1970.

Middleton, J. T. "Air Quality Criteria-Scientific Cornerstones of the 1967 Air Quality Act," Journal of Occupational Medicine, 10:9 (September, 1968), 535-38.

_____, and W. Ott. "Air Pollution and Transportation," Traffic Quarterly (April, 1968), 175-89.

Mills, C. W. The Power Elite. New York: Oxford University Press, 1956.

Molotch, H. "Oil in Santa Barbara and Power in America," Sociological Inquiry, 40 (Winter, 1970), 131-44.

_____. "Toward a More Human Human Ecology," Land Economics, 43:3 (August, 1967), 336-41.

Moynihan, D. "The Moynihan Report." The Negro Family: The Case for National Action. Washington, D.C.: Office of Planning and Research, U. S. Department of Labor, March, 1965.

Murphy, C. F. Governing Nature. Chicago: Quadrangle Books, 1967.

Nourse, H. O. "The Effects of Air Pollution on House Values," Land Economics (May, 1967), 181-89.

Ogburn, W. F. "Cultural Lag as Theory." William F. Ogburn on Culture and Social Change. Otis D. Duncan, ed. Chicago: Phoenix Books, 1964. Chapter 7. Reprinted from Sociology and Social Research, (January-February, 1957).

_____. Social Change: With Respect to Culture and Original Nature. New York: B. W. Huebsch, 1922, pp. 200-13.

Park, R. "The Urban Community as a Spatial Pattern and a Moral Order." Human Communities. New York: Free Press, 1952.

_____, E. Burgess, and R. D. McKenzie, eds., The City. Chicago: University of Chicago Press, 1925.

Prindle, R. A. "The Health Aspects of the Urban Environment," Ekistics, 25:151 (June, 1968), 428-31.

Public Awareness and Concern with Air Pollution in the St. Louis Metropolitan Area. U.S. Department of Health, Education, and Welfare, Public Health Service, Washington, D.C.: May, 1965.

Purdom, P. W. "The Environmental Health Challenge: Environmental Health in the Metropolitan Setting," Journal of Health and Human Behavior, 5:4 (Winter, 1964), 136-41.

Quinn, J. A. "The Nature of Human Ecology: Reexamination and Redefinition," Social Forces, 18 (December, 1939), 161-68.

Ramo, S. Cure for Chaos: Fresh Solutions to Social Problems Through the Systems Approach. New York: McKay, 1969.

Rankin, R. E. Air Pollution and the Community Image. Terminal Report to the U.S. Public Health Service, 1968. Grant No. AP-00460-01.

_____. "Air Pollution Control and Public Apathy," Journal of the Air Pollution Control Association, (August, 1969), 565-69.

Ray, Paul H. "Human Ecology, Technology, and the Need for Social Planning," American Behavioral Science, 11:6 (July-August, 1968), 16-19.

Robbins, M. C. "Explorations in Psychocultural Bioclimatology." Ph.D. dissertation, University of Minnesota, 1966.

Rogers, Everett M. Social Change in Rural Society. New York: Appleton-Century, 1960.

Rossano, A. T., Jr., et al. "Nationwide Air Pollution Control Training Efforts: Status Quo vs. Needs," Journal of the Air Pollution Control Association, 18:3 (March, 1968), 180-82.

Rossi, Peter. "How Can We Get Action For Clean Air Through Public Communications?" Washington, D.C.: U. S. Public Health Service, 1963. Pub. No. 1022.

Rydell, C. P., and G. Schwartz. "Air Pollution and Urban Form: A Review of Current Literature," American Institute of Planners Journal (March, 1968), 115-20.

_____, et al. Air Pollution and Urban Population Distribution. New York: Urban Research Center, Hunter College, 1968.

Sacco, J. F., and E. C. Leduc. "An Analysis of State Air Pollution Control Expenditures," Journal of the Air Pollution Control Association, 19:6 (June, 1969), 416-19.

_____. Also see Political Science Section.

Savas, E. S. "Feedback Controls on Urban Air Pollution," IEEE Spectrum, 6:7 (July, 1969), 77-81.

Schattschneider, E. E. The Semisoverign People. New York: Holt, Rinehard, and Winston, 1960.

Schmid, C. F. "Land Values as an Ecological Index." Proceedings of the Pacific Sociological Society, 1940.

Schueneman, J. J. "Organizational and Operational Techniques for Relieving Manpower Shortages in Air Pollution Control Agencies," Journal of the Air Pollution Control Association, 17:10 (October, 1967), 670-72.

Schusky, J. See Political Science Section.

Selznick, P. TVA and the Grass Roots. Berkeley: University of California Press, 1949.

Sewell, W. R. D. "Emerging Problems in the Management of Atmospheric Resources: The Role of Social Science Research," Bulletin of the American Meterological Society, 49:4 (April 1968), 326-36.

Shiffman, M. A. "A Survey of the Use of Standards in the Administration of Environmental Pollution Control Programs," Ph.D. dissertation, University of Pennsylvania, 1967.

Siegel, G. B. "Research in the Social Sciences for Determining Strategies of Air Pollution Control Administration," Public Health Reports (December, 1967), 1101-4.

Silberman, C. E. Crisis in Black and White. New York: Random House, 1964. Especially ch. 3.

Sills, D. The Volunteers: Means and Ends in a National Organization. New York: Free Press, 1957.

Smith, W. S., et al. "Public Reaction to Air Pollution in Nashville, Tennessee," Journal of the Air Pollution Control Association, 14:10 (October, 1964), 418-23.

Special Report to the Regents. The Polluted Air: Smog Research at the University of California. Berkeley, California, 1967.

Stalker, W. W., and C. B. Robison. See Political Science Section.

Stanley, W. J., and A. N. Heller. See Planning Section.

State of California, Department of Public Health. "Air Pollution: Effects Reported by California Residents." The California Health Survey. Berkeley, undated but apparently 1956.

Stern, A. C., ed. Air Pollution. Vol. 3. New York: Academic Press, 1968.

Sussman, M. B., ed. Community Structure and Analysis. New York: Crowell, 1959.

"Technical and Social Problems of Air Pollution: Symposium." New York: Metropolitan Engineers Council of Air Resources, Engineering Foundation, 1966.

"This Ecology Craze," The New Republic, (March 7, 1970), 8-9.

Thomas, H. "The Animal Farm: A Mathematical Model for the Discussion of Social Standards for the Control of the Environment Quarterly Journal of Economics (February, 1966), 143-48.

U. S. Department of Health, Education, and Welfare, Public Health Service, Consumer Protection and Environmental Health Service, National Air Pollution Control Administration. Air Pollution Publications. Pub. No. 979. Washington, D.C., 1969.

_____. Air Quality Criteria for Particulate Matter. Pub. No. AP-49. Washington, D.C., January, 1969. pp. 99-102.

Van Arsdol, M. D., et al. "Reality and the Perceptions of Environmental Hazards," Journal of Health and Human Behavior, 5:4 (Winter, 1964), 144-53.

Werthman, C. S. "The Social Meaning of the Physical Environment," Ph.D. dissertation, University of California, Berkeley, 1968.

Wheeler, J. O. "Transportation Problems in Negro Ghettos," Sociology and Social Research, 53:2 (January, 1969), 171-79.

Williams, J. D., and F. L. Bunyard. Interstate Air Pollution Study: Phase II Project Report: Opinion Survey and Air Quality Statistical Relationships. Cincinnati: Robert A. Taft Sanitary Engineering Center, May, 1966.

Wohlers, H. C. et al. "Can Air Pollution Be Controlled by Legislation?" Scientia (January-February, 1969), 58-64.

GENERAL

Abelson, P. H., "Progress in Abating Air Pollution," Science, 167 (March 20, 1970).

"Air Conservation for the Metropolis: The Need for Cooperation," International Affairs, 32 (May, 1964), 13-15.

"Air Pollution Bibliography," Air, and Water Pollution International Journal, Oxford, 10 (September, 1966), 641-46.

"Air Pollution Control: An Action Plan and Bibliography," American County Government, 32 (February, 1967), 29-40.

"Air Pollution Control in British Power Plants," World Power Conference, Section C1, Paper 82, Moscow, 1968.

"Air Pollution in the Schools and its Effects on Our Children," Consumer Bulletin, 52:24 (May, 1969).

"Air Pollution Publications: A Selected Bibliography." Washington, D.C.: Department of Health, Education, and Welfare, Public Health Service, Pub. No. 979.

"Air Pollution Titles," Guide to Air Pollution Literature, 2:6 (November-December, 1966), University Park, Pa., Pennsylvania State University.

"Allies of Smog," Economist, 217 (December 4, 1965) 1076.

Anderson, D. O. "The Effects of Air Contamination on Health," Canadian Medical Association Journal, Toronto, P. 1, 97 (September 2, 1967); P. 2, 97 (September 9, 1967), 585-93; P. 3, 97 (September 23, 1967), 802-6.

Automobile and Air Pollution, A Program for Progress, Parts One and Two. Washington, D.C.: U. S. Department of Commerce, 1967.

Ayres, R., and A. Kneese. "Environmental Pollution," Federal Programs for the Development of Human Resources. Pt. 2. Washington, D.C.: U. S. Government Printing Office, 1968, pp. 626-36.

Barringer, A. R. "We Could Perish in Our Heat," Financial Post, 62:55 (September 21, 1968).

"Bibliography on Air Pollution," Highway Research Record, 278 (1969), 12-14.

Burgdorf, L. P., and I. E. Harney. "Air and Water Pollution: Ideal for Classroom Study," Instructor, 79 (August, 1969), 118-19.

Cady, Cullison. "For Cleaner Air: The Massive Effort of the Motor Industry, the Petroleum Industry, and Government to Control Automobile Emissions: Part I," Highway User (November, 1967), 14-17.

Carr, A. B. "Suggestions for Teaching About Air Pollution," School Science and Math, 64 (March, 1964), 229-35.

Chase, F. R. "What's in the Air? Uncontrolled Dust, Fumes and Gases that Circulate Throughout the School," Safety Education 44 (February, 1965), 14-16.

Cooper, A. G. "Air Pollution Publications, a Selected Bibliography, 1963-1966." Washington, D.C.: Public Health Service, Pub. No. 979, 1966.

Crocker, T. Timothy, John R. Goldsmith, et al. "Human Health Effects Task Force Assessment." Task Force No. 2, Vol. 2. Riverside: University of California, Project Clean Air, 1970.

Darling, F. F., ed., Future Environments of North America. New York: Natural History Press, 1967.

Dasmann, R. F. Destruction of California. New York: MacMillan, 1965.

Densham, W. E. "Want Clean Air?" International Affairs, 33:8 (April, 1965).

"Detroit Says Smog is Licked," Steel, 169:26 (June 30, 1969), 7-8.

Devos, A., N. Pearson, P. L. Silveston, and W. R. Dryan. The Pollution Reader. Quebec: Harvest House, 1968.

Dubos, R. "The Human Environment," Science Journal, 5A:4 (October, 1969), 75-80.

Eleen, J. W. "Stop Pollution at Its Source." Canadian Lab., 12 (June, 1967), 12-15, 33-34.

Englund, H. M. "A Bibliography of Air Pollution Films," Journal of the Air Pollution Control Association, 17:5 (May, 1967), 281-86.

"Environmental Degradation and Education," Nation's Schools, 85:4 (April, 1970), 53-68.

Erickson. "Meteorological, Topographical and Geographical Factors Influencing Air Pollution." European Conference on Air Pollution, 1964. pp. 147-63.

Ewald, W. R., Jr. Environment for Man: The Next Fifty Years. Bloomington: Indiana University Press, 1967.

Farber, S. M., and R. H. Wilson, eds. The Air We Breathe: A Study of Man and His Environment. Springfield, Ill.: Thomas, 1961.

Ferrand, E. "Urban Air," Science Technology, 90 (June, 1969), 8-16.

Final Report, Select Committee on Air Pollution and Smoke Control. Toronto: Legislative Assembly, Province of Ontario, February 14, 1957.

Garnett, A. "Some Climatological Problems in Urban Geography with Special Reference to Air Pollution," Institute of British Geographers Transactions, 42 (1967), 21-43.

Goldsmith, J. R. "Los Angeles Smog," Science Journal, 5 (March, 1969), 44-49.

Gougan, R. C. "Clearing California's Skies: Concerted Campaigns to reduce Air Pollution Undertaken by State, Local and Federal Agencies," Public Utilities Fortnightly, 76 (September 30, 1965), 11-18.

Governmental Air Pollution Agencies, Directory, 1956-1964. Published annually by the Air Pollution Control Association, Pittsburgh, Pennsylvania, in cooperation with the U.S. Department of Health, Education, and Welfare, Public Health Service, Division of Air Pollution, Washington, D.C.

Grantham, J. N. "Keeping the [U.S.] Air Clean," Foreign Trade, 131 (January 4, 1969), 27-28.

Green, Leon, Jr. "Energy Needs Versus Environmental Pollution: a Reconciliation?" Science, 156 (1967), 1448-50.

Hayakawa, K. "Education for Air Pollution," Japanese Air Cleaning Association, Tokyo, 3:1 (1965), 4-7.

Heller, A. N., J. J. Schueneman, and J.D. Williams. "The Air Management Concept," Journal of the Air Pollution Control Association, 16 (June, 1966), 307-9.

Hoelscher, H. E., W. R. Turkes, and J. I. Abrams. "Educators Have Vital Role in Environmental Engineering," Environmental Science Technology, 3 (March, 1969), 235-40.

Jarrett, H., ed. Environmental Quality in a Growing Economy. Baltimore: Resources for the Future, Inc., John Hopkins Press, 1965.

Johnson, C. C., Jr. "Environmental Control-Challenge and Opportunity," Food, Drug, Cosmetic Law Journal, 24 (December, 1969), 568.

Kennedy, R. F. "Air Pollution and the Death of Our Cities," Social Action, 34 (May, 1968), 38-46.

Kirov, N. Y. "Some Effluent Problems of our Affluent Society," Clean Air, 2 (September, 1968), 3.

Leighton, Phillip A. "Geographical Aspects of Air Pollution," Geographical Review, 56:2 (April, 1966), 151-74.

Lewis, H. R. With Every Breath You Take. New York: Crown, 1965.

Marshall, J. The Air We Live In. New York: Coward-McCann, Inc.

McHugh, E. W. "The Effect of Rapid Transit on San Francisco Bay Air Quality," Journal of the Air Pollution Control Association, 17 (May, 1967), 277-79.

Meetham, A. R. Atmospheric Pollution: Its Origins and Prevention. New York: Pergamon Press, 1964.

Megonnell, W. H. "Air Pollution Control: Its Impact Upon Municipalities, Industry, and the Individual," Air Engineering, 9 (July, 1967), 12-14.

Metzler, D. F. "States Cooperate to Fight Air Pollution," Royal Society Health Journal, London, 88 (March-April, 1968), 75-78.

Middleton, J. T. "Middleton Talks About . . . ," Environment, Science, Technology, 2 (October, 1968), 734-37.

Miller, N. I. Index to Air Pollution Research, A Guide to Nonprofit and Industry Supported Air Pollution Research. University Park, Pa.: Information Office, Center for Air Environment Studies, July, 1967.

Moroz, W. J. "Learning to Live with Pollution," Executive, 8 (May, 1966).

Morse, A. L., and J. C. Juergensmeyer, "Air Pollution Control in Indiana in 1968: A Comment," Valparaiso University Law Review, 2 (Spring, 1968), 296-314.

Morse, H. N. "Air Pollution," Journal of the American Medical Association, 204 (June 24, 1968), 213-14.

"Nailing Detroit with the Blame for Foul Air," Production Engineer, 40 (February. 10, 1969), 17-18.

1969 Directory of Governmental Air Pollution Agencies. Pittsburgh: Air Pollution Control Association, 1969.

"Nonsense and Niggardliness Are Undermining Pollution Control," Modern Manufacturing, 1 (September, 1968), 184-86, 190.

Olds, F. C. "Air Pollution Control: Good Intentions in Search of Direction," Power Engineering, 73 (September, 1969), 28-35.

Perry, J. *Our Polluted World: Can Man Survive?* New York: Franklin Watts, Inc., 1967.

"Pollution: Everybody's Adversary: Symposium," *Today's Health*, 44 (March, 1966), 37-65.

Proceedings of the Mid-Atlantic States Section, Air Pollution Control Association, Semi-Annual Technical Conference. Philadelphia, March, 1966.

Proceedings of the Third National Conference on Air Pollution, December 12-14, 1966. Washington, D.C.: Public Health Service, Pub. No. 1949, 1967.

Progress in the Prevention and Control of Air Pollution: Second Report to the Congress of the United States, 1969. U. S. Secretary of Health, Education, and Welfare, 81st Congress, 1st Session, Senate. Doc. 91-11.

Puglisi, E. A., and W. R. Loftus. "Air Pollution, Incineration, and the Construction Industry: Sources of the Problem and Potential for Solution," *Construction Review* (September, 1969), 4-7; and (October, 1969), 4-12.

Purdom, P. W. "The Environmental Health Challenge: Environmental Health in the Metropolitan Setting," *Journal of Health and Human Behavior*, 5, 136-41.

Putnam, R. G., and F. J. Taylor. "Air Pollution--A Major Problem," *World Affairs*, 33 (January, 1968), 23.

Reitze, A. W., Jr. "Pollution Control: Why Has it Failed?" *ABAJ*, 55 (October, 1969), 923.

Rich, T. A. "Air Pollution Studies Aided by Over-all Air Pollution Index," *Environmental Science Technology*, 1 (October, 1967), 796-800.

Richardson, M. S. "Bibliography on Air and Water Pollution," *Special Librarian*, 57(August, 1966), 385-90.

Rienow, R., and L. T. Rienow. *Moment in the Sun*. New York: Dial, 1967.

Rogers, S. M. "A Brief Review of the Public Health Service Air Pollution Program," *Journal of the Pollution Control Association* 12 (July, 1962), 318-21.

Savas, E. S. "Feedback Controls on Urban Air Pollution," IEEE Spectrum, 6 (July, 1969), 467.

Scorer, R. S. Air Pollution. New York: Pergamon Press, 1968.

Slade, D. H. "Modelling Air Pollution in Washington, D.C., to Boston Megalopolis," Science, 157, 1304-7.

Smith, R. G. "An Analyst's View of Our Polluted Planet," Analytical Chemistry, 40:7 (June, 1968), 24a-32a.

Sparrow, C. J. "A Survey of New Zealand Air Pollution Literature and a Bibliography," Public Health, 83:2 (1968), 25-33.

State and Local Programs in Air Pollution Control. Washington, D.C.: Public Health Service, Pub. No. 1549.

"States are Key in Pollution Abatement," Environmental Science Technology, 3 (March, 1969), 223-28.

Stephens, Edgar R., Monty A. Price, and Joseph V. Behar. "Assessment of the Role of Peroxybenzoyl Nitrate in Eye Irritation in Photo, Chemical Smog." Research Project 5-18, Vol. 4. Riverside: University of California, Project Clean Air, 1970.

Stepp, J. M., and H. H. Macaulay. "The Pollution Problem," Special Analysis, (October 10, 1968).

Stern, A. C., ed. Air Pollution. Three Volumes. New York: Academic Press, 1968.

_____. "Present Status of Atmospheric Pollution in U. S.," American Journal of Public Health, 50 (March, 1960), 355-56.

Sussman, V. H. "What Progress Has Been Made in Air Pollution?" Pennsylvanian (January, 1968), 16-18.

Symposium: Air Over Cities. Washington, D.C.: Public Health Service, R. A. Taft Sanitary Engineering Center, Tech. Report, A62-5, 1967.

Tabershaw, I. B. "A New Health Professional: The Environtologist," Archives of Environmental Health, 18 (May, 1969), 811-15.

Tabor, E. C. "Contamination of Urban Air Through the Use of Insecticides," New York Academy of Sciences Transactions, 28 (1966), 569-78.

Technical Advisory Committee, Recommended Ambient Air Quality. Standards, A Report to the California Air Resources Board by the Technical Advisory Committee, September, 1970.

Truhaut, R. "Danger Thresholds," European Conference on Air Pollution (1964), 81-118.

"Urban Air Pollution with Particular Reference to Motor Vehicles." World Health Organization, Technical Report Service, No. 410, 1969.

"Urban Ecology and the Air Environment, Plenary Session, Summary," Journal of the Air Pollution Control Association, 19 (November, 1969), 836-46.

Williams, J. D., and N. G. Edmisten. "Air Resource Management Plans for the Nashville Metropolitan Areas." Washington, D.C.: Public Health Service, Pub. No. 999-AP-18, 1965, pp. 137-42.

Wise, W. Killer Smog, the World's Worst Air Pollution Disaster. Chicago: Rand McNally.

Wolman, Abel. "Air Pollution: Time for Appraisal," Science, 159: 3822, 1437-40.

——————. "Pollution as an International Issue," Foreign Affairs, 47:1 (October, 1968), 164-75.

Zimmer, C. E., and G. J. Noble. "The Impact of Computers Upon Air Pollution Research," Journal of the Air Pollution Control Association, 18 (June, 1968), 383-86.

ABOUT THE EDITOR AND THE CONTRIBUTORS

PAUL B. DOWNING is currently an Assistant Professor of Economics at the University of California, Riverside. Prior to this appointment, Dr. Downing served in the U.S. Army as an economist in the Office of the Assistant Secretary of Defense for Systems Analysis, where he was in charge of regional impact analysis of defense expenditures in the United States.

Dr. Downing studied at the University of Wisconsin, Milwaukee and received a doctorate in economics from the University of Wisconsin, Madison. In 1965 he was the recipient of a Public Health Service Fellowship to study Water Resources Management.

ROBERT J. ANDERSON is an Assistant Professor of Economics at Purdue University. He received his Ph.D. from the University of Pennsylvania in 1969. His publications have appeared in journals such as American Economic Review and Journal of Regional Science.

THOMAS D. CROCKER is currently an Acting Associate Professor of Economics at the University of California, Riverside. He is also serving as an Acting Research Associate for the Statewide Air Pollution Research Center at the University. Dr. Crocker has contributed his services to the Technical Committee of Southeastern Wisconsin Citizens for Clean Air, and he also holds the position of Chairman of the TR-3 Committee on Economic Effects of the Air Pollution Control Association. His primary area of research involves the optimal property right configurations with respect to the attributes of property objects. Dr. Crocker received his Ph.D. from the University of Missouri, Columbia.

ROSS CHARLES FOLLETT is a graduate student at the University of California, Santa Barbara, where he is studying sociology.

MICHAEL L. FOX is a graduate student of economics at the University of California, Riverside.

JAMES E. KRIER is an Acting Professor of Law at the University of California in Los Angeles. He received his degree in law from the University of Wisconsin Law School in 1966. Professor Krier is on the Advisory Board of the Environmental Law Reporter, a faculty advisor to Project Clean Air, and a member of its Social Science Task Force. He is also a member of the working committee for the Environmental Science and Engineering Program at UCLA, and a co-principal investigator of UCLA's Ford Foundation Program in Planning and Environmental Control Law.

ROBERT O. LOVERIDGE is an Associate Professor of Psychology and an Associate Dean at the University of California, Santa Barbara. Dr. Loveridge received his Ph.D. in 1965 from Stanford University. He is a member of both the American Political Science Association and the Riverside City Charter Revision Commission. In 1970, Dr. Loveridge received the University of Pacific Alumni Award. He has published articles in both the areas of sociology and psychology.

HARVEY L. MOLOTCH is presently a Visiting Associate Professor at the State University of New York at Stoney Brook. He holds a Ph.D., which he received from the University of Chicago in 1968. He has contributed his work to such publications as American Journal of Sociology, Land Economics, and Sociological Inquiry.

ROBERT W. REYNOLDS is an Associate Professor of Psychology at the University of California, Santa Barbara, where his duties include both undergraduate and graduate instruction. Dr. Reynolds has done extensive research in the area of psychology and he has contributed to such journals as, Psychological Review, American Psychologist, and Science. Dr. Reynolds holds a Ph.D. in psychology from the University of Buffalo.

Soc
HC
110
A4
A65

DUE DATE			
~~NOV 2 9 1991~~			~~MAR 0 6 2004~~
~~DEC~~ ~~1992~~			
~~FEB 1 8 1993~~			
~~JUN 0 4 1996~~			
~~APR 3 0 1996~~			~~DEC 1 1993~~
			Printed in USA